William Morris Fontaine, Israel Charles White

The Permian or Upper Carboniferous Flora

Of West Virginia and S.W. Pennsylvania

William Morris Fontaine, Israel Charles White

The Permian or Upper Carboniferous Flora
Of West Virginia and S.W. Pennsylvania

ISBN/EAN: 9783337063207

Printed in Europe, USA, Canada, Australia, Japan

Cover: Foto ©berggeist007 / pixelio.de

More available books at **www.hansebooks.com**

(SECOND GEOLOGICAL SURVEY OF PENNSYLVANIA:
REPORT OF PROGRESS
PP.)

THE PERMIAN OR

UPPER CARBONIFEROUS FLORA

OF

WEST VIRGINIA

AND

S. W. PENNSYLVANIA.

BY

WM. M. FONTAINE. M. A.,

Late Professor of Chemistry and Physics in the University of West Virginia.
Now Cocoran Professor of Geology in the University of Virginia.

AND

I. C. WHITE, A. M.,

Professor of Natural History in the University of West Virginia, and Assistant
Geologist on the Geological Survey of Pennsylvania.

WITH THIRTY-EIGHT PLATES.

HARRISBURG:
PUBLISHED BY THE BOARD OF COMMISSIONERS
FOR THE SECOND GEOLOGICAL SURVEY.
1880.

BOARD OF COMMISSIONERS.

1880.

ASSISTANT GEOLOGISTS.

Persifor Frazer, Jr.—Geologist in charge of the Survey of Adams, York, Lancaster and Chester counties.

Ambrose E. Lehman—Topographical Assistant, for mapping the South Mountain.

Frederick Prime, Jr.—Geologist in charge of the Survey of Northampton, Lehigh, and Berks counties.

E. V. d'Invilliers—Topographical Assistant, for mapping the Easton-Reading range.

Franklin Platt—Geologist in charge of the Survey of the coal fields of Tioga, Bradford, Potter, Lycoming, and Sullivan counties.

W. G. Platt—Geologist in charge of the Survey of Armstrong and Jefferson counties.

R. H. Sanders—Topographical Assistant in Dauphin, Lebanon and Berks counties.

I. C. White—Geologist in charge of the Survey of Crawford and Erie counties.

J. F. Carll—Geologist in charge of the Survey of the Oil Regions.

H. M. Chance—Geologist in charge of the Survey of Clinton and Clarion counties.

C. A. Ashburner—Geologist in charge of the Survey of McKean, Elk, Cameron and Forest counties.

A. W. Sheafer—Assistant in McKean county, &c.

F. A. Genth—Mineralogist and Chemist at Philadelphia.

F. A. Genth, Jr—Aid in the Laboratory.

A. S. McCreath—Chemist, in charge of the Laboratory of the Survey, 223 Market street, Harrisburg.

John M. Stinson—Aid in the Laboratory.

C. E. Hall—Geologist in charge of the Survey of Delaware county, and Palæontologist in charge of the Museum.

N. A. Stockton—Aid in the Museum.

M. G. Carraher—Aid in the Museum.

Charles Allen—Assistant in locating outcrops in Delaware county, and for Records of Railroad and other Levels, Harrisburg.

H. C. Lewis—Volunteer geologist for the survey of the gravel deposits of south-eastern Pennsylvania.

Leo Lesquereux—Fossil Botanist, Columbus, Ohio.

E. B. Harden—Topographer in charge of Office Work, &c. 1008 Clinton street, Philadelphia.

O. B. Harden—Assistant in preparing illustrations.

F. W. Forman—Clerk in charge of the Publications of the Survey, 223 Market street, Harrisburg.

TABLE OF CONTENTS, PP.

Page.

Preface by the authors, ... vii

INTRODUCTORY CHAPTER.

Vespertine group, (Pocono formation,) 3
    Its flora, ... 6
Umbral shale group, (Mauch Chunk formation,) 9
Conglomerate group, (Pottsville formation.) 10
    Its flora, ... 11
Lower productive coal measures, ... 15
    Their flora, ... 16
    Horizon of the Kittanning coal bed, ... 17
Lower barren measures, ... 19
    Their flora, ... 20
Upper productive coal measures, ... 21
Upper barren measures, ... 24
    Three general sections, ... 29

CHAPTER 2.

Description of Species.

Equisetites, ... 33
Calamites, ... 34
Sphenophyllum, ... 36
Annularia, ... 38
Sphenopteris, ... 40
Neuropteris, ... 46
Odontopteris, ... 52
Callipteris, ... 54
Callipteridium, ... 55
Pecopteris, ... 61

	Page.
Goniopteris,	81
Cymoglossa,	84
Alethopteris,	87
Tæniopteris,	90
Rhacophyllum,	93
Caulopteris,	95
Sigillarea,	96
Cordaites,	97
Rhabdocarpus,	98
Carpolithes,	98
Guiliemites,	99
Saportæa,	99
Baiera,	103
Gerablattina,	104
Description of plates,	121
Index to names of species,	135

PREFACE.

About two years ago a number of well preserved plant impressions were observed at Cassville, in Monongalia County in West Virginia, in the shale associated with the coal bed worked at that village. This bed is called the "Waynesburg," in the nomenclature used in describing the strata of the "Appalachian Coal Field." It is the highest worked bed which occupies any important area in this great field.

The beauty of the impressions, and the fact that some of them were new, led to further examination. The result was, that many new forms were found at this place; other and remote localities were visited, and it was found that at many exposures the coal was accompanied by remarkably fine plants. Various horizons above this bed were found to afford plant impressions, many of them being new. Where these horizons are exposed in the adjoining States of Pennsylvania and Ohio, they appear to afford comparatively few plants. There was little prospect then, that the surveys of these States would obtain material sufficient to throw much light on the plant life of the upper beds of this important coal field. There was also little prospect that the State of West Virginia would, in any short time, authorize such a survey as would give this material to the scientific world.

As the result of our collections seemed to us to be of some interest and value to science, we were induced, in our private capacity, to study this material, and prepare our conclusions for publication as a contribution to science.

We were led to confine our examinations and collections to the strata above the Pittsburg coal bed by several inducements. One important reason was, that only in this portion of the Carboniferous strata could we expect to find any change in the flora pointing to the assumption of a Permian facies. Again, the fullest development of the highest beds occurs in West Virginia, and seem to be richer in plant impressions than elsewhere. Another reason was, that the plant impressions found in the Pittsburg and underlying beds, had already received, or would soon receive, ample study in the surveys of the adjoining States.

Unfortunately we met with many difficulties in our examinations and collections. The interval between the Pittsburg coal and the Waynesburg is almost barren of plants. This is due, in part, to their destruction by maceration, and in part, to the fact that much of the interval is occupied by limestones and other rocks deposited under water during an extensive submergence. The paucity of material from this series of strata is all the more to be regretted seeing that the period during which it was deposited seems to be marked by important changes in the flora.

The Waynesburg coal, as noted elsewhere, is exceptionally rich in plants, and is the highest horizon where they can be obtained with tolerable ease, or in any abundance. This bed, when it is exposed, is generally worked, and thus affords access to undecomposed shales yielding well preserved plants. But even in this bed, as the plants occur in the shales left by the miners as a roof, special excavations had to be made by us to gain the plants.

Owing to persistent search, and to visits paid to every point promising good material, we can claim to have made a collection from this bed which is fairly representative of its flora. Some of the localities which have afforded us good material from this bed are as much as 70 miles in air line distance apart, and the points examined are quite numerous.

Above the Waynesburg bed the exposures are few and poor. Few excavations have been made for railroads and other public works. No coal bed is worked on this horizon. The strata are mostly of a material which soon breaks down into clay and loose matters which obscure the outcrops. Beside this, there has been much surface action, and owing to the depth to which decomposition has penetrated it is very difficult to get sound material. These causes have prevented us from procuring from the uppermost beds such a collection as we wished; hence we cannot claim to present, from this portion of the field, such comprehensive details as from the Waynesburg coal bed. Still, we have plants from widely separated localities and horizons in this upper portion, and they suffice to give important indications of the character of its flora.

As we made our collections ourselves we had opportunities to study the plants *in situ*, and to examine and compare a large amount of material which we could not have had if the collecting had been done by others.

In the preparation of our work we are indebted to Prof. Newberry, of Columbia College, New York, for the use of books, and especially to Prof. Lesquereux of Columbus, Ohio, for the liberal loan of many works not otherwise accessible to us.

Mr. G. Gutenberg and Prof. Mertz, of Wheeling, W. Va. have kindly loaned to us, for the purpose of comparison, specimens of plants collected by them from the horizon of the Pittsburg coal, near Wheeling. Mr. T. L. Hazzard, of Washington and Jefferson College, Penn., also loaned us plants collected by him from the Upper Barrens of Pennsylvania. To all these gentlemen we return our thanks.

THE AUTHORS.

W. VA. UNIVERSITY,
MORGANTOWN, W. VA., *June 25, 1878.*

NOTE.—The order of our names on the title page has no significance, as we are equally and jointly responsible for this work.

THE AUTHORS.

THE FOSSIL FLORA

OF THE

UPPER CARBONIFEROUS OR PERMIAN

IN

WEST VIRGINIA

AND

SOUTH-WEST PENNSYLVANIA.

INTRODUCTORY CHAPTER.

A sketch of the geology of the Carboniferous Formation in West Virginia.

The different groups which include the strata of Carboniferous age, as found in the Appalachian Coal Field, are in some respects better defined, and more sharply separated from each other in West Virginia than elsewhere.

The names finally adopted by the former State Geologists of Virginia and Pennsylvania, Prof. W. B. and H. D. Rogers, for the grand divisions of the series, commencing with the lowest, were Vespertine Sandstone, Umbral Limestone, Umbral Red Shale, Seral Conglomerate and [Productive] Coal Measures.

These had been previously numbered by them X (Vespertine), XI (Umbral), XII (Conglomerate) and XIII (Coal Measures.)

In the Reports of Progress of the Second Survey of Pennsylvania the following names have been employed:—Pocono Sandstone (X), Mountain Limestone, Mauch Chunk Red Shales (XI), Pottsville Conglomerate (XII), Coal Measures (XIII.)

Where the Catskill (IX) or "Old Red Sandstone" of the English Geologists is found, it is capped by the Vespertine, but where it is wanting, the Vespertine succeeds the Chemung (VIII.)

The Vespertine forms a well marked horizon, including the oldest of the strata of Carboniferous age. The Umbral, or No. XI of the old surveys, includes two very different series. The lower portion is a great limestone, corresponding to the "Mountain," or "Lower Carboniferous Limestone," of European Geologists. The upper portion is, along the eastern border of the Virginia field, a mass of shales and sandstones, mostly red. Owing to the fact, that in Pennsylvania, the limestone thins out, and gives place to a great mass of shales and sandstones, some geologists class the limestone as a member of the Umbral Group. As, however, in West Virginia, the limestone is in great force, and is well distinguished from the overlying sandstones and shales, and as it is identified with a world-wide formation, we will place it as an independent member, in the grand divisions.

The Productive Coal Measures are found to be naturally divided into four well marked groups, two yielding workable beds of coal, and two, which are almost devoid of persistent workable coal beds. Accordingly, the fifth division above named, or the Productive Coal Measures, has been sub-divided into (a) The Lower Productive Measures, (b) The Lower Barren Measures, (c) The Upper Productive Measures, (d) The Upper Barren Measures.

In this sketch of the general geology of our field, we do not pretend to give the reader anything more than a mere outline, sufficient to enable one not familiar with the order of succession, and the character of the strata, to form an intelligent idea of the occurrence of the plants, and to gain some knowledge of the changes shown, in ascending from lower to higher horizons, in the Carboniferous strata of West Virginia.

The Vespertine Group.

At the base of the Carboniferous strata in West Virginia, we find the Vespertine Group, which contains the oldest land plants yet found in this State. The group being a shore formation, is quite variable in thickness, thinning as we proceed from the east (which was the ancient shore) to the west. It thickens in a remarkable manner to the north and south of the northeastern corner of the State. Along the eastern border of this State, it ranges in thickness from about 200 feet, to over 1000 feet. The lower limit is given on the authority of Prof. Wm. B. Rogers, who determined this thickness on the Potomac River, in Hampshire county.

Owing to the absence of fossils, and the great variability of the strata for some distance above the highest strata of the Chemung containing fossils, there is some difficulty about fixing the base of the Vespertine, unless, as seems natural, we accept as a base the first stratum which, by its persistence and well marked physical character denotes a decided and widely prevailing change in the conditions of deposition. Such a stratum we find in a peculiar conglomerate which everywhere makes one of the lower members of the Vespertine Group, and which is the very lowest which can be identified at widely separated points.

This rock is a highly siliceous white sandstone, and is almost always a pronounced conglomerate. It usually has large pebbles which have the peculiarity of being flat, instead of possessing an ovoid or elliptical form, as is usually the case with the pebbles of conglomerates. These flat pebbles characterize this rock at widely separated points, such as Montgomery Co. Va. and Cheat River in Monongalia Co. W. Va.

Between this conglomerate and lower beds which are certainly Chemung, as shown by their fossils, there is an interval occupied by strata which contain no fossils. The lowest of these have a bright red color and the general character of the Catskill group as seen elsewhere, and hence they may be of that age. The upper beds, lying next under the conglomerate above described, are mainly flaggy sandstones, of a dingy grey color, and may be of Vespertine age. At the White Sulphur Springs in Greenbrier Co. W. Va. the red beds are about 340 feet thick, and the flaggy sandstones occupy an interval of about 500 feet.

Where all the members of the Vespertine exist in Virginia the group is triple, composed of the conglomerate and firm silicious sandstones at the base, with a middle portion of grey sandy shales containing coal, and an upper member of red shales and sandstones. All three vary a good deal in thickness, but the coal of the middle member is usually found included within an interval of 100 feet. Two distinct beds are usually found, about 40 feet apart, but the coal is sometimes found distributed in thin layers a few inches thick over a space of 40 or 50 feet. This is the composition of the group along the eastern and southeastern border of W. Virginia. In Montgomery Co. Va. we find two workable beds of coal in the middle member. At the White Sulphur, these may still be distinguished, but they have thinned down to 6 or 8 inches. In W. Va. along the southeastern border, there are no persistent workable beds, and the coal exhibits a tendency to break up into thin layers. The coal is found near the central portion of the middle member. The overlying red rocks, forming the third member, are by some geologists considered as forming a portion of the Umbral and placed in one group with the Lower Carboniferous Limestone. But the limestone forms a clearly defined limit to these rocks, and there was evidently an important change, at this horizon, in the condition of the deposition. In the absence of all fossils from these strata, it would seem best, in W. Va. at least, to place them with the underlying Vespertines, into which they pass by insensible gradations.

As stated above, Prof. Wm. B. Rogers found the entire

group only 200 feet thick on the Potomac River, near the villages of Westernport, in Hampshire Co. W. Va. From this point of minimum development it thickens to the northeast, southeast, and west, but much more rapidly in the two directions first named. Towards the northeast, in Huntingdon Co., Pennsylvania, according to Mr. Ashburner, it is 2133 feet thick, with a coal bearing member near its center, 303 feet thick, which contains 19 small seams of coal.*

Traced to the southeast, in Montgomery Co. Va. we find it nearly 2700 feet thick, and containing, as previously stated, two important coal beds.

While the group, as a rule, becomes much thinner as we follow it to the west, yet traced in this direction from Westernport, the locality of its least development, it thickens. Hence in the eastern part of Monongalia Co. W. Va. on Cheat River, as recently determined by us, it is over 500 feet thick, the base not being seen.

The physical character also changes, as shown in the western exposures, for we find on Cheat River the following strata in descending order :

1. Lower Carboniferous (Umbral) Limestone.
2. Flaggy Sandstones. 90 feet.
3. Massive White Sandstone. 100 feet.

[* That is, 2133 feet up to the base of the *red beds beneath the Mountain Limestone*. Mr. Ashburner very properly excluded these *red beds* from the Vespertine, and considered them the lower member of the Umbral. I cannot agree with Professors Fontaine and White in thinking that " it seems best in W. Virginia " or any where else " to place them with the underlying Vespertines into which they " certainly do not, at least along an outcrop of 150 miles in Pennsylvania, " pass by insensible gradations."

The Mountain Limestone is an interpolated deposit in the red shales, since it thins away to nothing in eastern Pennsylvania ; in Middle Pennsylvania not only lies 141 feet above the well marked lower limit of the *red shales*, but it is itself nothing but a group of frequently alternating *red shales, red shaly limestones, red silicious limestones, variegated *red and grey* limestones, *red and grey* mottled calcareous shales, &c., through a vertical space of 45 feet. There seems to me no more reason for making the *Mountain Limestone* of XI a horizon line separating two great formation, than for using the *Ferriferous Limestone* of the Lower Productive Coal Measures, or the *Great Limestone* of the Upper Productive Coal Measures for that purpose. At all events, any such line of demarkation would be absurd for the nomenclature of our Anthracite Coal Region.—J. P. L.]

4. Flaggy Sandstones. 300 feet.

5. Conglomerate with flat pebbles not fully exposed.

No. 2 is conglomeratic in many portions. The base of No. 5 was not seen. No coal was seen here in the group, and, so far as observed, it is always wanting in the more westerly outcrops.

The Vespertine Flora.

The collections made of Vespertine plants, are rather meagre, hence caution must be used in making deductions from the material obtained. The number of localities, however, affording plants is considerable, and as they are widely separated we may consider that the facts observed, have considerable weight in fixing the character of the flora.

In the first place, we are struck by the distinctly characterized facies, which would, in every collection made from any locality, at once indicate its Vespertine age. Indeed most of the plants do not pass above this group. Such as do are cosmopolitan forms of wide vertical and horizontal range.

Another noteworthy feature is that while the number of individuals of a species at a given locality is often very great, the number of species is small, and we find one or two plants forming the entire flora. The most abundant species found at the localities in W. Va. are the following :

Lepidodendron Veltheimianum. Sternb.

L. . . . . Sternbergii, Brongt.

Triphyllopteris Lescuriana, (Cyclopteris Lesc. of Meek.)

T. . . . . Virginiana, (C. . . . Virg. of Meek.)

Archaeopteris (Cyclopteris, Daws. Noeggerathia, Lesq.) obtusa. Lesq.

A. Alleghanensis.(Cyclopteris All. of Meek.)

A. (Noeggerathia Bock. Lesq.) Bockschiana, Goep.

A. . . (Palaeopteris Hib. Schimp.) Hibernica. Parb.

Besides these, we find commonly, several species of Lepidodendron allied to Veltheimianum ; several species of Archaeopteris of the type of A. Jacksoni, (Cyclopteris Jacksoni, Daws.) ; one or more species of Triphyllopteris, all not described as yet.

More rarely we find a Neuropteris allied to N. flexuosa, but, if not identical with it, a plant allied to Dawson's Cyclopteris valida; Cardiopteris frondosa, Schimp. and other plants.

The localities yielding the most abundant plants are Lewis Tunnel near the White Sulphur Springs; the Dora Coal Field in Augusta Co. Va.; and the coal beds of Montgomery Co. Va. These are points whose extreme distance apart is more than 100 miles. The Lewis Tunnel locality yields the greatest variety of plants.

The plants which especially characterize the group are the Lepidodendra, the Archaeopterids, and the Triphyllopterids.

The Triphyllopterids form two types. The first has the lobes less deeply cut, but broad and obtuse, like Meek's Cyclopteris Virginiana, or Dawson's Cyclopteris valida. The second type, has narrow, deep, and pointed lobes, like Meek's Cyclopteris Lescuriana. There are probably several new species of each kind; but in the case of these plants, and of the Archaeopterids, the transition from one form to another is so gradual that a large amount of material is needed to establish new species.

The Archaeopterids also show two types. That which is most abundant in species and individuals has narrow and small pinnules, like Dawson's Cyclopteris Jacksoni. The second type has broader and more flabellate leaflets, like the Noeggerathia obtusa of Lesquereux.

There is apparently a transition, on the one hand through the type of Archaeopteris Jacksoni into the Triphyllopterid form, with broad obtuse lobes, and on the other hand through the type of Archaeopteris obtusa into the form of Cardiopteris.

Indeed all these plants, as well as the broad leaved Sphenopterids of the lower coals, such as Sphenopteris macilenta, have the facies of Archaeopteris.

Besides these positive features, there is a negative one, which, of course, so long as the collections are meagre, cannot possess much weight.

No Pecopterids, Sphenopterids, Neuropterids (with one

exception,) and no Sigillariae, not to mention more recent forms, have as yet been found, and this deficiency adds much to the antique aspect of the flora.

From Pennsylvania Mr. Ashburner gives, on the authority of Prof. Leo Lesquereux, the following species in his " Measured Section of the Paleozoic Formations :"

Sphenopteris flaccida.
Ulodendron majus. L. & H.
Stigmatocanna Wolkmanniana.
Knorria acicularis. Goepp.
Stigmaria minuta. Goepp.
Lepidodendron. Spec ?

The Umbral, or Lower Carboniferous Limestone.

The only fossils found in this limestone are invertebrate, and they show that it corresponds in age with the Lower Carboniferous or Mountain Limestone. In West Virginia it is a well defined and thick mass ; but in Pennsylvania the limestone thins out almost entirely, while the red shales and sandstones, which in W. Virginia, mainly overlie it, become greatly developed. The same condition of things appears to exist to the southeast, in Montgomery Co. Va. This passage of the limestone into the shales and sandstones of the Umbral causes a difficulty in the grouping of the Umbral and the Limestone, and has led some geologists to place both in one group. On the other hand, to the west and southwest the shales and sandstones disappear, and leave the limestone with increased thickness. As showing the variations in thickness of this rock we give the following measurements :

Near the White Sulphur Springs in Pocahontas Co. Prof. Wm. B. Rogers determined its thickness to be 822 feet. Towards the north it thins rapidly, for near Westernport Prof. Rogers found it only 80 feet thick. On Cheat River, in Monongalia Co. it is about 100 feet thick, and 25 miles farther north, in Fayette Co. Pa. it is according to Stevenson only 40 feet thick. In Huntingdon Co. Pa. Mr. Ashburner finds it to be 49 feet thick.*

[* See foot note to page 5 above ; and Report of Progress Second Geol. Sur. Penna. F, 1878, page 195.—J. P. L.]

This group, so far as known, contains in W. Va. no fossil plants. The subsidence causing the deposition of this limestone, and the accompanying destruction of plant life no doubt had an important influence in bringing about the change which we find to have taken place in the flora of the Conglomerate Series, which is the next plant-bearing horizon above the Vespertine.

The Umbral Shale Group.

This group, in West Virginia, consists of shales and sandstones of various hues and textures, and, where fully developed, like the Vespertine group, possesses a tripple character. The lower member consists of red, poorly laminated shales or marlites, and red or brown, argillaceous sandstones. The shales are remarkable for their deep blood-red color and crumbling, friable texture. The middle portion is mainly composed of pretty siliceous sandstones, of a grey or white color ; grey flags ; and grey or greenish marlites. The upper member is, like the lower, composed of deep red shales and sandstones. On New River, in the vicinity of Richmond Falls, the group appears in great force. Here it is nearly 1500 feet thick. It shows at this locality the triple division in a marked manner. This group also is a shore formation, and reaches its greatest development in the east, thinning out entirely as we pass to the west. Like the Vespertine, it shows great variations in thickness, even along the eastern border, and follows nearly the same law of change.

The following measurements will indicate the variations in different quarters :

Prof. Wm. B. Rogers finds it in Pocohontas Co. near the White Sulphur to be about 1310 feet thick, while on the Potomac River near Westernport, Md. he finds it to be 738 feet thick.

In Huntingdon Co. Pa. Mr. Ashburner finds 1100 feet of Umbral Rocks, including 49 feet of Umbral or Lower Carboniferous Limestone, in several layers.† On Cheat River in Monongalia Co. the interval between the Lower Carbon-

[† And including also, in his 1100 feet, 141 feet of reddish and green Umbral shales below the Limestone.—J. P. L.]

iferous Limestone and the Conglomerate Group is occupied by sandy shales 170 feet thick. These are of a grey color, and contain no red material, except one or two feet of red crumbling marlite immediately in contact with the limestone.

No plants have been found in the Umbral in W. Va. and no important coal beds are known to exist in it. In the western part of Greenbrier Co. and near Quinnimont in Fayette Co. W. Va. one, and perhaps several coal beds exist near the top of the group. This portion of the State is but little explored, and may yield plants.

The Conglomerate Group.

This group also, where fully developed in West Virginia, forms a triple series.

The typical arrangement is as follows: At the base we find a massive conglomerate, often of brownish grey color. In the center, shales and flaggy sandstones, containing coal beds, alternate with massive siliceous sandstones. At the top we have a heavy bedded white siliceous sandstone, with many conglomerate layers.

The upper bed is the most persistent of the series, and forms the floor of Coal Measures of West Virginia.

The Conglomerate, like the groups above described, varies much in thickness and composition. This is especially true of the middle and lower members. The middle, or coal-bearing member, often thins out so as to bring the upper and lower members close to each other, and then, the coal is almost, or quite, cut out. The lower member is often wanting, as in East Tennessee, and possibly in Alabama.

The character of the strata, and of the coal beds, indicates rather rapid subsidence, and frequent sudden changes in the conditions of deposition. As a consequence we find the coal beds varying rapidly in thickness, even when workable, but usually too thin to be of much value. The variable character of the beds underlying the upper member causes them to contrast strongly with the more uniform strata found above it, which constitute that portion of the Carboniferous Formation commonly called "The Productive Coal Measures."

The following measurements will indicate the character of the group at different points :

At Quinnimont, on New River, in Fayette county, West Virginia, it perhaps attains its maximum development. Here. at the base, we find a conglomerate. 80 feet thick ; in the middle, a great series of shales and sandstones, with nine coal beds. This middle member is about 950 feet thick, and is overlaid by a massive sandstone, largely conglomeratic, 150 to 200 feet thick. The coals are mostly thin and variable. Only one bed is known to be workable over an extended area.

The same group continues south into East Tennessee, and probably into Alabama.

Farther north, in Randolph county, West Virginia, Dr. Stevenson finds it 600 feet thick, with at least one coal bed, near the central portion.

At the northern line of the State, on Cheat River, in Monongalia county, the group is 325 feet thick. The top is a massive sandstone, highly conglomeratic, 175 feet thick. Under this, at some localities, a small coal bed, with some associated shales is found, and at these places the entire interval between the coal and the base is occupied by sandstone, similar to that lying above. The coal is not persistent, for at other localities it is wanting, and the entire group is composed of massive conglomeratic sandstone.

Flora of the Conglomerate Group.

At numerous points where the shales associated with the coal beds are exposed, we find many well preserved plants, and the sandstones yield great numbers of nut-like fruits. On New River. at Quinnimont. and at Sewell station, we find the following plants :

Alethopteris Helenæ, Lesqx.
A. . . . . . lonchitica, Brt. Var.
A. . . . . . grandifolia, Newb.
Sphenopteris Hœninghausi, Brt.
S. . . . . . obtusiloba, Brt.
S. . . . . . macilenta, L. & H.
S. . . . . . adiantoides, L & H.
Lepidodendron selaginoides, Sternb.
Calamites cannæformis, Schloth.

Pecopteris nervosa, Brt.
P. . . . muricata. Brt.
Neuropteris Smithiana, Lesqx.
N. . . tenuifolia, Brt.
Megalopteris Hartii. Andr.
M. . . Sewellensis, Font.
Odontopteris neuropteroides, Newb.
O. . . . gracillima, Newb.
Asterophyllites acicularis, Daws.

Besides the above named, we find a small Archæopteris, very near to Dawson's Cyclopteris Jacksoni; a Cordaites, allied to C. Robbii, Daws; and fragments of what must have been a very large leaf resembling a Tæniopteris. The midrib of this is broad, and from it, closely placed, parallel nerves pass at right angles. The facies of this plant resembles closely Tæniopteris Smithii, Lesqx. from the lower coal beds of Alabama, and also the genus Orthogoniopteris, of Andrews, founded on plants occurring in the lower coal strata of Ohio. The above list does not assume to be exhaustive of the plants found on New River, in the Conglomerate. It may be stated here, as *Megalopteris Sewellensis* has never been figured, that it is a plant near Neuropteris (Megalopteris) Dawsoni, as given by Dawson, but the leaflets are smaller, thicker, and not so acuminate as in this plant.

Alethopteris Helenæ, Neuropteris Smithiana, and Tæniopteris Smithii, are figured and described in Prof. Lesquereux's Report P on the "Coal Flora of Pennsylvania, &c." 1879.

In Western Pennsylvania the most abundant plants of this group are:

Alethopteris lonchitica, Brt.
A . . . . . . grandifolia, Newb.
Neuropteris flexuosa, Brt.

Pecopteris nervosa, Brt.
Sphenopteris macilenta, L. & H.

Besides these, numerous species of Lepidodendron, Sigillaria and Cordaites are found, with a great number of fruits belonging to the Genera Trigonocarpus, Cardiocarpus, and Rhabdocarpus.

Professor Leo Lesquereux has given in "The Report of Progress of the Geological Survey of Alabama," for 1875, a list of plants sent to him from the coal field of Alabama, by Dr. Smith, the chief of the Survey of that State. In this there are many plants identical with those found in the Conglomerate on New River, and some which occur only at these two localities. Sphenopteris Hæninghausi, Brt., Neuropteris Smithiana, Lesqx., and Alethopteris Helenæ, Lesqx. are plants common to the two localities, and not found elsewhere in the Appalachian Coal Field. Pecop-

teris muricata, Brt. and P. nervosa, Brt. are abundant in
the Conglomerate on New River, as well as in Alabama.
Besides these, we find as common to both localities Sphenopteris obtusiloba, Brt., Alethopteris lonchitica, Brt., and
others.

It may not be possible to establish by stratigraphy the
existence of the Conglomerate Group in Alabama, but the
identity of many of the plants, and the close resemblance
of the facies of the flora found on New River and in Alabama point strongly to the Conglomerate age of at least
the lower portion of the Alabama Coals.

If Mr. Richard P. Rothwell is correct in his report on
"Alabama Coal and Iron," quoted by Dr. Smith in the
above mentioned Report of Progress, the stratigraphy also
indicates the existence of the Conglomerate Group, for he
mentions two groups separated by a Conglomerate, and
states that the lower one contains 8 coal beds.

One of us has had recently an opportunity to examine
a collection of plants made from the lower coals of East
Tennessee, and he found the species identical with those
existing in the Conglomerate on New River. We may then
conclude that a portion of the coal beds of this State are
also of Conglomerate age.

There is a remarkable resemblance between the Conglomerate flora as determined in West Virginia, and that of the
lower coals of Ohio, up to coal No. 4, as given by Dr. Newberry and Prof. Andrews. With few exceptions the plants
are identical, and the general facies of both floras differs
from that of the Productive Coal Measures. The finding
of plants in the Conglomerate of W. Va. similar to those
of Prof. Andrews, such as Archœopteris and Megalopteris,
along with many of the species occurring with Coal No. 1 of
Ohio, seems to indicate no great difference in the age of the
three floras. Dr. Newberry states that in Ohio the flora of
Coal No. 1 is characteristic, that it changes with Coal No.
4, and that above the latter no divisions can be made in the
plants.

Prof. Lesquereux gives Whittleseya elegans as found in
Alabama, and as it occurs nowhere else, except in the flora

of Coal No. 1 of Ohio, it is a very significant bond of union between this bed and the Alabama Coals.

Taken as a whole, the flora of the conglomerate group has a well characterized facies which distinguishes it from that of the Vespertine below, and from that of the Productive Coal Measures above. It retains some of the Vespertine types, in the Archaeopterids, and possibly the Megalopterids, (though the latter have not as yet been found in the Vespertine of W. Va., but in Canada are Devonian.) It possesses a large number of plants peculiar to itself, or not found above it. Among these we may mention the large coarse Alethopterids, A. grandifolia and several varieties of the A. lonchitica, along with the typical form; the peculiar Odontopteris neuropteroides; Neuropteris Smithiana, and many others. Again it possesses a considerable quota of plants which, with specific changes, pass up into the Productive Coals.

Alethopteris lonchitica, and its varieties, is a plant highly characteristic of the group. The Pecopterids are few, and in the case of the P. muricata, and P. nervosa, which are perhaps the most abundant, show composite types, including the features of the true Pecopterids, with those of the Neuropterids, and Sphenopterids. These not fully differentiated forms find their analogues in the composite type shown in a group of Sphenopterids, which is especially characteristic of the Conglomerate flora. This group, including Sphenopteris macilenta, L. & H.; S. obtusiloba, Brt.; S. latifolia, Brt., and others, retains the facies of the obtusely lobed Triphyllopterids, in conjunction with features marking the true Sphenopterids, and Pecopterids.

The Productive Coal Measures.

This as a whole, is distinguished from the Conglomerate group, by the greater uniformity of the conditions under which the various strata and Coal beds, were formed. As stated in another connection, this group is naturally divided into sub-groups, each of which requires a separate description. We will commence with the lowest of these:

The Lower Productive Coal Measures.

This series of strata is limited below by the upper member of the Conglomerate, and above by the Mahoning Sandstone. This latter, is usually a thick sandstone, often conglomeratic, and forms a natural base to the next series above, viz: The Lower Barren Measures.

The Lower Productive Coal Measures, like the group last described, attains its maximum thickness in the southern part of the State, and thins greatly in passing north. It has its greatest development along the Great Kanawha River, in Kanawha, and the adjoining counties, where it is not less than 1200 feet thick. The details of the geology of this portion of the State, are not known. No minute examinations in the interests of pure science, have ever been made here. The investigations which are made, are usually in behalf of land-owners, or purchasers, and have for their object the determination of the number, character, thickness, &c., of the coal, and iron-ore beds. Enough however is known of the stratigraphy, to show that the 1200 feet of rocks are almost entirely devoid of limestone, but are composed of thick strata of shale, and sandstone, holding numerous, valuable beds of coal.

Mr. M. F. Maury, M. E., has made a section of the Lower Productive Measures, on Paint Creek, Kanawha Co. at a point where the base of the series is not shown. Yet in this section, 974 feet of strata are shown, holding 14 coal beds, whose united thickness is 51 feet 10 inches, besides 7 beds, whose out-crop only, was seen.

The flora of this portion of the Productive Measures, is entirely unknown. From the accounts given by amateur collectors, it would seem to be abundant and varied.

In the northern portion of the State, both the stratigraphy, and the character of the flora, are better known, though our knowledge of the latter, is still imperfect. Here the entire thickness of the series is barely 250 feet. We find more limestone, with fewer and thinner beds of coal. There are only 4 important coals, aggregating about

15 feet in thickness, and of these only two are workable over large areas.[*]

Flora of The Lower Productive Measures.

No special search has been made for plants in this portion of the Coal Strata in West Virginia, and no doubt the list given below might be largely increased by further investigations. Two horizons have yielded most of the plants. The lowest is that of the Kittanning Coal Seam near the base of the Series, and the highest is that of the Upper Freeport Coal Seam near the top.

From the Kittanning Coal we have:

Neuropteris heterophylla. Brt.	Lepidostrobus ornatus. L. & H.
N. . . . . . Clarksoni Lesq.	Lepidophyllum. Spec?
Lepidodendron Sternbergii. Brt.	

From the Upper Freeport we have:

Neuropteris acutifolia. Brt.	Pecopteris arborescens. Schloth.
Odontopteris subcuneata. Bunb.	Asterophyllites rigidus. Brt.

At both horizons the following plants occur:

Neuropteris flexuosa. Brt.	Pecopteris villosa. Brt.
N. . . . . . hirsuta. Lesqx.	Sphenophyllum Schlotheimii. Brt.
N. . . . . . rarinorvis. Bunb.	

But in Western Pennsylvania Mr. I. F. Mansfield has made a large collection of plants for the Second Geological Survey of Pennsylvania from the Darlington bed, which next overlies the Kittanning bed; and Prof. Lesquereux, the fossil botanist of the Survey, has determined from this material the following species, published in Report of Progress Q, White, 1878, p. 55.

[* Considering the known thickness of the Lower Productive Coal Measures, "barely 250 feet" in the northern counties of West Virginia,—considering that this thickness is wonderfully well preserved in Pennsylvania for a hundred miles north north-west into the Beaver Valley country, and for more than 150 miles north north-east nearly to the New York State line,—and considering the absence of reliable data for identification in Middle and Southern West Virginia, acknowledged in the text,—one cannot be too cautious in generalizing respecting so extraordinary a thickening of the series in that direction. My own surveys on Sandy waters in East Kentucky in 1864, led me to quite the opposite view; for the normal thickness is maintained in that region, if the Hill Sand Rock of Tug Fork be the Mahoning. It will need much "minute examination in the interests of pure science" between the Cheat and the Kanawha before the Mahoning Sandstone can be rightly placed on the latter river; and until that be done it is unsafe to dogmatize about the thickening of the *Lower* and thinning of the *Upper Coal Measures* in that direction.—J. P. L.]

Fossil Plants from the Horizon of the Killanning Coal.

CALAMARIA.

Asterophyllites:
equisetiformis.
foliosus.
sublævis.

Calamites:
approximatus.
Suckowii.
ramosus.
nodosus.

Sphenophyllum:
Schlotheimii.
longifolium.
emarginatum.

Annularia:
Sphenophylloides.
longifolia.

Equisitites:
infundibuliformis.

Calamostachys:
tuberculata.

FILICES.

Cyclopteris:
trichomanoides.
obliqua.
elegans.
undans.
fimbriata.

Neuropteris:
angustifolia.
cordifolia.
hirsuta.
Clarksoni.
flexuosa.
tenuifolia.
vermicularis.
plicata.
Loschii.
crenulata.

Odontopteris:
Schlotheimii.

Dictyopteris:
obliqua.

Callipteridium:
Mansfieldi.

2 PP.

Alethopteris:
ambigua.
lonchitica.
Serlii.
Sullivantii.
nervosa.
Pluckeneti.

Pecopteris:
hemiteloides.
microphylla.
truncata.
Sillimani.
squamosa.
plumosa.
polymorpha.
chærophylloides.

Sphenopteris:
Newberryi.
mixta.
artemisiæfolia.

Hymenophyllites::
lactuca.
laceratus.
Gutbierianus.
expansus.

Spiropteris:
villosa.

Stemmatopteris:
Mansfieldi.

Caulopteris:
obtecta.

LYCOPODIACEÆ.

Lepidodendron:
obovatum.
Sternbergii.
quadratum.
modulatum.

Lepidophyllum:
undulatum.
Mansfieldi.
auriculatum.
foliaceum.

Lepidostrobus:
ornatus.
variabilis.

Lepidophloios:
 laricinus.

SIGILLARIÆ.

Sigillaria:
 monostigma.
 alternans.
 reniformis.
 mamillaris.
 sculpta.
 elliptica.
 tessellata.

Syringodendron:
 pes-caprioli.
 cyclostigma.

Stigmaria:
 ficoides.

Cordaites:
 borassifolia.
 principalis.
 Mansfieldi.
 reflexa.

Dicranophyllum:
 species.

Cordianthus:
 fl. masculina (1 Species.)
 fl. femina (Antholithes) 2 Species.

Artisia:
 transversa.

FRUCTUS.

Carpolithes:
 vesicularis.
 multistriatus.
 platimarginatus.
 clypeiformis.
 fraxiniformis.
 Canneltoni.

Rhabdocarpus:
 Boeckschianus.
 clavatus.
 amygdalæformis.

Trigonocarpus:
 Daviesii.

Cardiocarpus:
 mamillatus.

RADICES.

Pinnularia:
 capillacea.

FUNGI.

Rhizomorpha:
 sigillariæ.

Since the publication of the above list, Prof. Lesquereux has published in the " Proceedings of the American Philosophical Society," a paper on Cordaites, in which he gives the following additional species:

Cordaites:
 validus.
 crassus.
 grandifolius.
 communis.
 diversifolius.
 gracilis.
 costatus.
 serpens.
Cordianthus:
 gemmifer.
Cordaistrobus:
 Grand Euryi.

Dicranophyllum:
 dimorphum.
Taeniophyllum:
 deflexum.
 contextum.
 decurrens.
Desmiophyllum:
 gracile.
Lepidoxylon:
 anomalum.

The above lists give the plants found at one locality only, and though this occurs in Pennsylvania, the plants may be considered as representing also the flora of the Lower Productive Measures of West Virginia. Of course, with more extended and careful search, we may expect to find many additional species. The lists are especially valuable, as showing the change which has taken place in the grouping of the plants since the Conglomerate period.

The Lower Barren Measures.

This series takes its name from the comparatively small amount of workable coal which it contains. It has for its base the Mahoning Sandstone, and extends up to the Pittsburg Coal bed. Its thickness, in the southern part of the State, is not known, but is perhaps about 700 feet. It is the last of the groups which have their maximum thickness in the South. In the northern portion of the State, its thickness ranges from 550 to 600 feet.

Its physical character is pretty uniform. The base is composed of a sandstone, (the Mahoning,) which is usually thick and coarse, and quite often conglomeratic. From near the base to the middle portion we find some thin marine limestones. One of them, the highest persistent limestone showing marine fossils, is noteworthy as being the last stratum which gives evidence of the extensive prevalence of marine conditions, and for its great extent and uniform character. Though hardly ever more than two feet thick, it extends over an area in W. Virginia of more than 30,000 square miles, showing everywhere the same lithological character, and containing the same fossils.

This stratum, the "Crinoidal Limestone," of the Ohio and Pennsylvania surveys, is of great importance, as a geological horizon, since it furnishes an easily recognized initial plane.

Up to this horizon, the incursions of the sea were not uncommon, as is shown by the marine fossils of the limestones of the underlying groups. Limestones are not uncommon in the succeeding measures above, but they are usually impure, and of fresh water origin. There must then, at this point, have been an important change in the physical geography of the coal field.

Associated with this limestone, and passing higher in the series, we find incoherent shales, of a brilliant red, or mottled color. These alternate with grey sandstones and shales, and are covered at the top of the series by impure fresh water limestones.

Flora of the Lower Barren Measures.

There are but few horizons in these measures which afford plants, and but little examination of these has been made. The most promising. is that about 20 feet below the Pittsburg coal. This horizon, in the vicinity of Wheeling, West Virginia. has yielded the following plants:

Neuropteris :
  hirsuta, Lesqx.
  rarinervis, Bunb.
  acutifolia, Brongt.
  flexuosa, Brongt.
  Loschii, Brongt.
  Grangeri, Brongt.
Sphenopteris :
  furcata, Brongt.
  minuti-secta, Sp. nov.
Pecopteris :
  Pluckeneti, Brongt.
  Bucklandi, Brongt.
  (Alethopteris, Lesqx.,) spinulosa.
  Candolleana, Brongt.
  notata, Lesqx.
  dentata, (plumosa form,) Brongt.
  pteroides, Brongt.
Alethopteris :
  aquilina, Brongt.
  Sp. nov. allied to A. Gigas of Geinitz.

Lescuropteris :
  Moorii. (Lesq.); (Sch.)
Odontopteris :
  Sp. nov. allied to obtusiloba of Naumann.
Annularia :
  longifolia, Brongt.
  sphenophylloides, Ung.
Cordaites :
  borassifolius, Ung.
Sphenophyllum :
  filiculme, Lesqx.
  trifoliatum, Lesqx.
Asterophyllites :
  Sp.? near equisetiformis.
Rhacophyllum :
  filiciforme, Schimp.
Calamites :
  cannaeformis, Schloth.
Syringodendron :
  pes-capreoli, Gein.

The above list is the result of but slight effort at collecting from this plant-bearing horizon, and could be largely increased by further search.

A specimen of Neuropteris hirsuta from this locality, shows six pinnules arranged as they would stand when attached to a common rachis, which unfortunately has been broken off from the stone. The locality is remarkable for the number of fruiting specimens of Pecopterids. Several fruiting leaflets of even Neuropteris hirsuta are found.

Many fine fruiting specimens of Alethopteris aquilina occur. There are numerous specimens of Pecopteris Candolleana, which differ somewhat from the forms found in the Waynesburg Coal at West Union, and which will be described further on. The plant at the horizon now in question has thinner leaflets, on which the nerves are very distinctly shown, while the West Union plant has very obscure nerves, and a very thick leaf-substance, as well as longer and more deciduous pinnæ. The remarkable plant, Lescuropteris Moorii, hitherto found only at a higher horizon, in the Upper Productive Measures, is found in detached pinnæ here.

The Upper Productive Coal Measures.

This is the only one of the sub-divisions of the Carboniferous Formation which has not a great sandstone everywhere at its base. But even in this case, we often find a tendency in the rocks of the Lower Barren Measures to pass into sandstone, within a short distance below the Pittsburg Coal bed.

The Upper Productive Coal Measures begin with the great Pittsburg Coal Bed, and end with the Waynesburg Coal. In the northern part of the State, the average thickness is about 350 feet. The series thus begins and ends with an important coal bed. The Pittsburg Coal, which forms the base, is the most widely extended and important coal bed in the Appalachian Coal Fields. It covers an area of more than 20,000 square miles in W. Virginia. Its greatest thickness is towards the east, where it is often from 10 to 14 feet thick, as is shown in Mineral Co. W. Va. and in the Cumberland Coal basin of Maryland. The least thickness is found in the southern part of the State, where, towards the southern line of its outcrop, it thins down to 3 or 3½ feet of coal.

But little is known of the character of the Upper Productive Coal Measures in the southern part of the State, but it seems evident that they are less developed there than in the northern portion, both in thickness, and in the number of the coal beds which they contain. In the south, we find but two beds in the series. The most important of these is the Pittsburg, which, in some places, attains a maximum thickness of 6 feet. The other coal lies above, at an unknown distance. Its thickness is not known, as it seems to be too unimportant to have attracted any attention. The comparatively small development of this series, in the south, is but a continuation of that change in the conditions controlling the deposition of the strata, which we find commenced in the underlying Lower Barren Measures, and which we will find intensified in the succeeding Upper Barren Measures. This change consists in the reversal of the comparative thickness of the groups in the northern and southern portions of the State, and in the production of a greater development to the northwards.

In the north, we find two coal beds, separated by small intervals from the Pittsburg. The lowest of these is the Redstone, which occurs 25 to 40 feet above it, and the other is the Sewickley, which is found 80 to 100 feet above. A third coal, not so persistent in W. Va. as the two last named, is found from 90 to 100 feet below the top of the series. This is the Uniontown Coal, a seam which attains its maximum development in the adjoining portions of Pennsylvania.

The strata composing the Upper Productive Measures in the northern portion of W. Virginia are limestones, often quite impure, grey shales, and argillaceous thinly bedded sandstones. The entire mass indicates the deposition of sediment in pretty deep water, during a widespread submergence which removed the shore lines to a considerable distance. Indeed the prevalence of fine sediment which marked the subsidence following the formation of the Mahoning Sandstone in the Lower Barren Measures, holds throughout the entire interval up to the Waynesburg Sandstone, and seems not to have been affected by the elevation of the surface which gave rise to the formation of the Pittsburg Coal. The great amount of limestone found in the interval between the Pittsburg Coal and the Waynesburg indicates a very considerable subsidence of the Appalachian Region where such a mass of limestone is found. This subsidence is of importance in furnishing a cause for the great difference shown in the flora of the Pittsburg and Waynesburg Coals. In order to bring out more distinctly this feature we give three graphic sections of the Upper Productive Measures, as found at three points, which may be taken as fairly representative of the whole.*

It must be noted that these limestones show no marine fossils, and none of the shells so abundant in the limestones up to the middle of the Lower Barren Measures are found in them. The only organic remains which they contain are a few minute bivalve crustaceans. They vary a good deal in composition, but are usually impure. They are most probably of fresh-water origin. A portion of them may have been formed in brackish water. These features all indicate that an important change in the physical features of the country took place towards the close of the period in which the Lower Barrens were formed.

We see from the sections, that in Monongalia Co. W. Va. we get between the Pittsburg and Waynesburg Coals 88 feet of limestone; at Wheeling, W. Va., not including the intercalated shales, 150 feet; and in Greene Co., Penn., 119 feet.

According to the geologist's method of reckoning time the formation of so much limestone, and of such a mass of fine shales, requires a long period, and this, combined with the amount of subsidence which must have occurred over wide areas, would fully explain that change in the facies of the flora which we find exhibited in the plants of the Waynesburg Coal bed. This change will be better understood after an examination of the fossils found associated with that coal seam.

The following section of the Waynesburg coal bed is given to show the mode of occurrence of the plants. The bed is one of the most important and persistent of the Upper Coal beds, covering as it does an area in West Virginia and Pennsylvania of at least 15,000 square miles. Where best developed it contains fully 8 feet of coal, and

---

* See Sections and Figs. 1, 2, 3, at the end of this chapter.

over large areas it is 5 or 6 feet thick, exclusive of partings.

Section of the Waynesburg Coal at Cassville, Monongalia Co., W. Va.

1. Roof shales, with many plants, ... 1 to 12 feet.
2. Coal, ... 12 inches.
3. Clay parting, with many plants, ...   6   "
4. Coal, ...   18   "
5. Shale of very variable thickness, ... 6 in. to 6 feet.
6. Coal, main layer, ...   4½ feet.
7. Floor.

The Upper Barren Measures.

These measures commence with the Waynesburg Sandstone (a rock which overlies the Waynesburg Coal bed) and
extends to the highest beds of the Carboniferous Formation. The existence of the Waynesburg Coal, and its accompanying sandstone, is not known in the Southern part
of W. Virginia. Hence the dividing plane between the two
measures is not made out in that quarter. Nothing is
known of the character and thickness of the Upper Barren Measures in that direction. Only their existence is
known, and the fact that they are much less developed
than in the north. Every indication from the few facts
known about this, and the series immediately underlying
it, points to the fact, that after the period of formation of
the Lower Barrens the area of great subsidence and abundant sedimentation was shifted from the southern portion
of the State to the northern.

The series along the northern line of the State, is much
better known, both in its stratigraphy, and its flora. Of
the plants of this and the preceding series in the south
we know absolutely nothing.

As these beds, to the top of the geological column, contain no fossils of consequence except plants, and as very
few of these have hitherto been collected and studied, the
entire mass of rocks, up to the highest exposures in the Appalachian Coal Fields, has been assumed to be of Carboniferous age rather from the lack of evidence to show the
presence of any other formation, than from any positive
proof that carboniferous strata do really extend to the summit of the column.

Whether any of these strata should be placed in a more recent series as, "Permo-Carboniferous," or "Permian," may be better determined after a review of the evidence afforded by the plant-life.

It is only necessary here to refer to the general section of the strata made in passing from the western part of Monongalia Co. where the highest strata occur, to the east where the Waynesburg Coal appears at Cassville.*

This section does not give the entire thickness of the Upper Barren Measures in W. Va., since in Wetzell and Marshall counties the column of rocks extends from 200 to 300 feet higher. We have had no opportunity to examine and measure these beds.

The upper 300 feet of the section given are never fully exposed, so that not much can be said about the strata occupying this space. A very massive sandstone is often found near the top, and probably one or two small limestones occur near the center, as they appear at this horizon in the adjoining portions of Pennsylvania.

In naming and numbering the different beds of limestone, coal, &c., found in this series, we have followed the nomenclature of Dr. John J. Stevenson in his Report of Progress, K, on Greene and Washington Counties in Pennsylvania, 1876.

No. 3 of the section was by him called Limestone X. It is one of the most persistent members of the series, as we have traced it over a wide area in Monongalia, Wetzel, and Marshall counties, always finding it at the proper horizon.

None of the coals of this series ever attain workable dimensions except No. 20, or the Washington Coal. The other coal beds are 1 to 1½ feet thick, and are never mined except by "stripping" at points where they lie near the surface.

The upper half of the series is quite variable in the character of its strata. In some places, we find it containing a great deal of massive sandstone, with drab, argillaceous beds, mainly incoherent shales. At other points, we find on the same horizon, several hundred feet of red shales,

---

*See Section and Fig. 4 at the end of this chapter.

often mottled with green, buff, or yellow spots, and streaks. Towards the south, the red and variegated shales increase in thickness, and descend lower in the series, sometimes even nearly to the horizon of the Waynesburg Coal. The red shales are quite conspicuous in Marshall Co., and in the 600 feet of strata shown at Bellton we find about 400 feet of red shales, not in a single bed, but in several beds, from 40 to 60 feet thick, alternating with brown sandstones or drab-colored shales.

The Waynesburg Sandstone, the rock which forms the base of the series, is an important stratum, since its physical character denotes plainly a great change in the conditions which had prevailed for a long period previous to the time of its formation. As has been previously stated, these conditions were quiet subsidence, and deposition of fine shales, with much limestone. But in the sandstone now described, we find many evidences of strong currents, which tore up the previously formed coal, and brought in a vast amount of coarse material. The approach of this unquiet condition of things is indicated in the structure of the Waynesburg Coal itself.

The Waynesburg Coal bed usually contains a parting of blue shale, near the middle, which shows extraordinary fluctuations in thickness. It sometimes disappears entirely, but rarely falls below 4 inches in thickness. The most common mode of occurrence is with fluctuations from a few inches up to several feet. It is not uncommon to find in a few yards distance, a sudden thickening from 5 or 6 inches up to six feet, and even more. The peculiarity of this shale is made more striking by the fact, that it possesses this character over an area of many thousand square miles; and while the changes are thus sudden the material is always of fine texture. It is to be observed that the plants yielded by this coal bed are found in the roof-shales, some distance above this variable parting.

The Waynesburg Sandstone, well characterized, forms a marked feature in the geology of the district where it occurs. Its usual thickness is from 50 to 75 feet, and its ordinary character that of a coarse conglomeratic rock, in which

the pebbles are often so numerous and large as to cause it
to rival the Great Conglomerate of the Coal Measures. The
rock is sometimes a mass of pebbles from $\frac{1}{4}$ of an inch to
one inch in diameter. This sandstone often descends, and
cuts out a portion, sometimes nearly all, of the underlying
coal. Immediately under it, come the roof shales of the
Waynesburg Coal, which contain the plant impressions
which form a considerable portion of those to be described
in the following pages.

The roof shales which yield the plants, are usually from
5 to 10 feet thick, of a dark dove color, and quite fine
grained. The plants are generally finely preserved. The
physical character of these shales is remarkably uniform,
and differs but little at widely separated localities, so that
it alone is sufficient to decide the horizon of the specimen
showing it, especially when containing some of the many
plants which it affords.

The most striking difference shown between the mode of
occurrence of the plants in the roof shales of the Waynesburg Coal and of those found *at higher horizons*, is seen in
the fact, that *in the latter* the species are few, while the
number of individuals is very great, and these species extend over the entire areas of the coal field. Thus we find
a few plants forming the entire flora of localities, when
from the immense number of individuals, and from the excellent preservation of the material, we are led to expect to
find a great variety.

At the Waynesburg horizon, on the contrary, while the
number of individuals of a species is great, we also find a
larger number of species. Again, we find the plants distributed in the most singular manner, they being grouped
in colonies, which are confined within very narrow limits;
so that the plants which abound in one opening for coal,
will be entirely wanting in another only a few hundred
yards distant, where we find instead of them a collection
of species so different, that it might well characterize a
different horizon. The same rule holds good at the exposures of the bed in other places, but not in so marked a
manner as at Cassville.

We find also at this place plants which have not as yet been seen at any other locality. The shales which contain the plants, though varying much in thickness, are pretty evenly bedded, and of fine texture, showing no evidence of differences of level or of currents which could account for the peculiar distribution of the plants. The unquiet condition of things marked by the fluctuations of the shale parting inclosed in the coal-bed, seems to have been succeeded by a period of quiet deposition of sediment, which was followed by an era of great disturbance, productive of the Waynesburg Sandstone.

Sections to illustrate the Introductory Chapter.

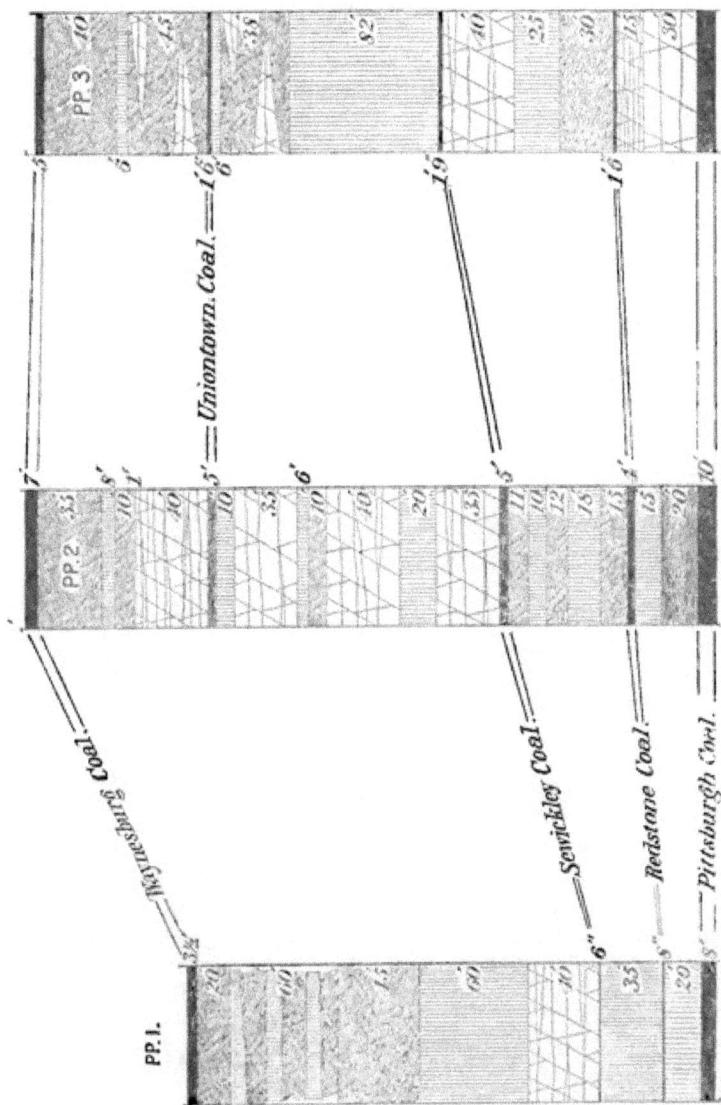

Fig. 1. *Upper Productive Coal Measures near Wheeling, W. Va.*

Waynesburg Coal,	3' 6"
Shales,	20'
Limestones and shales, interstratified,	60'
Shales,	45'
Limestone,	60'
Sandstone,	40'
Sewickley Coal,	6"
Sewickley Limestone,	35'
Redstone Coal,	8 '
Redstone Limestone,	20'
Pittsburgh Coal,	8'
	292' 8"

Fig. 2. *Upper Productive Coal Measures in Monongalia Co., W. Va.*

Waynesburg Coal,		7'
Shales,		35'
Limestone,		8'
Shales,		10'
Limestone,		1'
Sandstone, flaggy,		40'
(*Uniontown Coal*,) black slate,		5'
Limestone,		10'
Sandstone,		35'
Limestone,		6'
Shales,		10'
Sandstone,		40'
Limestone,		20'
Sandstone,		35'
Sewickley Coal,		5'
Shales,		11'
Sewickley	*limestone*,	10'
	shales,	12"
	limestone,	18'
Shales,		15
Redstone Coal,		4'
Redstone limestone,		15'
Slates and shales,		20'
Pittsburgh Coal,		10'
		382'

Fig. 3. Upper Productive Coal Measures, exposed near Rice's landing in Greene county. Pennsylvania.

Waynesburg Coal,	5'
Shales,	40'
Limestone,	6'
Sandstone and shale,	45
Uniontown Coal,	1' 6"
Uniontown limestone,	6'
Shale and sandstone,	38'
Great Limestone,	82'
Sewickley Coal,	1' 9'
Sandstone,	40'
Limestone,	25'
Shale, sandy,	30'
Redstone Coal,	1' 6"
Pittsburgh Upper Sandstone, { flaggy,	15'
{ massive,	30'
Pittsburgh Coal,	8'
	374' 9"

Fig. 4. Upper Barren Measures of Monongalia County.
W. Va.

Sandstones, shales, and concealed rocks,	300'
Sandstone, shaly, ...	50'
Limestone No. X, of Report K., (Stevenson,)	9
Shales, red, argillaceous, sandy and concealed, ...	170'
Washington Upper Limestone, ...	12
Shales and Sandstones, ...	40
Jolleytown Coal, ...	1' 6"
Shales, clays, and sandstones, ...	87'
—— *Coal,*	1'
Sandstones and shales, ...	37'
Washington Middle Limestone, ...	1'
Sandstones and shales, red, ...	70'
Coaly shales, ...	3
Shales and sandstone, ...	42'
Limestone, ...	2
Shales and sandstone, ...	35'
Plant bearing coaly shales, ...	3'
Washington Lower Limestone, ...	5'
Black slate, ...	4'
Washington Main Coal, ...	3
Sandstone, finely laminated, ...	18'
Washington Little Coal, ...	1'
Sandstone, massive, ...	18'
Waynesburg Coal B, ...	1'
Sandstone and shales, ...	30'
Limestone, ...	8'
Waynesburg Coal A, ...	1' 6"
Limestone, ...	1'
Shales, ...	5
Waynesburg Conglomeratic Sandstone, massive, ...	75
Plant bearing shale, ... 0' to 10	
	1,044'

PP.4.

DESCRIPTION OF SPECIES.

EQUISETIDES, Schimper.

Equisetides rugosus, Schimp.   Pl. I, Fig. 6.

In the roof shales of the Waynesburg Coal, at West Union, W. Va., we find fragments of a plant which seems more nearly allied to this species than any other.

The specimen best characterized is represented on Pl. I, Fig. 6.   It differs from the figure given by Geinitz and Schimper, in not seeming to be so fleshy, since the fragment represented in our figure seems to have been leaf-like in nature.   The base of the specimens seems to show that the fragment had been attached in a sheathing manner to a stem.

Equisetides elongatus, Sp. nov.   Pl. I, Figs. 1–4.

(Stem unknown, sheath comparatively very long, and wide, composed of cylindrical ribs, obtusely rounded at the end, and consolidated together; ribs fleshy, and marked with a cord of nerves, apparently composed of several vascular fibres united together; attachment apparently by the entire base, in a sheathing manner. Very deciduous.)

Fig. 1 shows a very long and broad sheath, whose attachment was not seen. Figs. 3 and 4 show the attachment, but do not extend up to the summit of the sheath. These sheaths must have been easily detached, since we looked carefully for the stems on which they might have been borne, but in a great number of specimens could only find the obscure attachments which we have figured. We find them in fragments, lying scattered through the shale. Some of these are even longer than the one figured in Fig. 1. On

3 PP.   (33)

one face of the ribs we see the nerve bundle distinctly marked, but the opposite side leaves in the shales only a smooth furrow-like impression, without any sign of nerves. This singular plant might at first sight seem only the impression of the stem of a calamite, but its fragmentary character, and the fact that the entire leaf substance is well preserved, with all its carbon intact, and that in this no trace of anything can be seen but the agglutinated rod-like ribs, precludes the idea of its being anything like a Calamite. The ribs which compose the entire plant, seem in their original condition to have been cylindrical, but they now appear flattened by pressure. Our plant resembles somewhat Goeppert's Bockschia flabellata, "Die Ton. Farnk." Pl. I, Figs. 1 and 2.

It seems allied in some respects to Phyllotheca, Brongt. and may stand as a connecting link between that genus and *Equisetides*. The sheaths seem to have been stripped off from the stem which bore them, in laminae, for we often find at the base, near what must have been the insertion, a thinning down of the sheath to a mere film of epidermal matter, as if it had been torn away from the stem which it had embraced.

Habitat—Roof shales of the Waynesburg Coal, West Union, W. Va.

Equisetides striatus, Sp. nov. Pl. I, Fig. 5.

(Stem unknown, sheath seen only in long narrow strips, formed of several consolidated, strongly striated ribs, which terminate in long slender teeth.)

The ribs show no central vascular cord, or mid-nerve, but are marked with very strong striae, which resemble nerves. This species is found sparingly, and never attached.

Habitat—Roof shales of the Waynesburg Coal, West Union, W. Va.

CALAMITES, Brongt.

This genus, so abundant at the lower horizons of the Carboniferous Formation, has almost disappeared at the higher levels. We find only one species, and that is very sparingly represented, only a few specimens being found in the entire mass of material at the localities which afford thousands of examples of other plants. Above the Waynesburg Coal we do not certainly find any Calamites.

Calamites *Suckowii*, Brongt.

This species is found very sparingly at the horizon of the Waynesburg Coal, at Cassville and West Union. It does not differ from the typical form, except perhaps in the greater flatness of the ribs.

Nematophyllum. gen. nov. (νημα, thread, φυλλον, leaf.)

Stem covered with a thick, very finely striate epidermis; internodes rather remote, swollen; leaves verticillate, numerous, very long and thread-like, of equal width throughout, finely striate, without nerves, united at the base, in a narrow annular band.

We have found it necessary to form a new genus to include the plant figured on Pl. II. Figs. 1, 2, 3, 4, 5, since it cannot properly be placed under *Asterophyllites*, (Calamocladus, Schimp) lacking some of the essential features of that genus, especially the ridged stem and leaves with mid rib. This genus is defined by all the authors as containing leaves free to the base, and furnished with mid rib. Heer, in his Pfl. d.. Steink. Per. d. Schw. p. 50, decribes a plant under the name of *Asterophyllites longifolius*, which would certainly not seem to be an Asterophyllites, but agrees closely with our genus. Again. Heer in Plf. d. Trias u. Jura. p. 78, describes under the name of *Schizoneura Meriani*, another species, which is not known to possess the essential feature of Schizonura, viz: union of the leaves at some stage of growth. This plant has nearly all the features of our genus, and most probably should be included in it.

Nematophyllum *angustum*. Sp. nov. Pl. II. Figs. 1–5.

The specific character of this plant is that of the genus, with the addition that the number of leaves is from 10 to 20.

their width 1½ to 2 mms., their length over 10 cms. The stem of the plant is usually from 1 to 1½ cms. wide, and is rather fleshy than woody in texture. Examined with a strong lens, the epidermis, as well as the leaves, show striæ which are quite distinct, and under the lens look like fine nerves. They are of the same kind in both, and the leaves contain usually from three to four.

We have seen leaves over 10 cms. long, and even then the ends were not preserved.

Habitat—Roof Shales of the Waynesburg Coal, Cassville and West Union, W. Va.

SPHENOPHYLLUM, Brongt.

This genus is well represented at the horizon of the Waynesburg Coal, both in species and in the number of individuals. It becomes very rare above this coal bed, and like a great many other genera, seems to have been almost extinguished during the formation of the conglomeratic sandstone which overlies the Waynesburg Coal. We have met with specimens only from one locality at the higher levels, and these were very few in number.

Sphenophyllum latifolium Sp. nov. Pl. I. Figs. 10 and 11.

Stem, rather strong and rigid, rough leaves, large and very broadly curvate, with the margin incised irregularly, forming lobes of unequal size, and irregular shape, lobes rounded dentate on the margins; nerves passing out flabellately from the insertion, and thrice forking, sending a branch into each tooth on the margin; whorls, composed of six leaflets, which are often more or less united near their insertion on the stem.

This plant is more nearly allied to Sphenophyllum longifolium, Germ., than any other described plant, but it differs from it in many important particulars, being wider, and not so long in proportion. We never find our plant with bifid leaves, a point which seems common in S. longifolium. The nervation also is quite different in the two.

The tendency of the leaflets to unite near their insertion is a feature peculiar to our plant.

Habitat—Roof Shales of the Waynesburg Coal, Cassville and West Union, West Virginia.

Sphenophyllum filiculmis, Lesqx., Pl. I, Fig. 8.

Fig. 3, Pl. I, represents *S. filiculmis*, Lesqx., "Geol. of Penn.," vol. II, part 2, Plate I, Fig. 6: It is one of the best characterized species of Sphenophyllum in the entire Carboniferous flora, always having its lower pair of leaflets shortened, and deflexed along the stem.

This plant is quite abundant at the horizon of the Waynesburg Coal. We find it in the roof shales of this bed at Cassville, West Union, and Carmichael's. It seems to be confined to this horizon, as we have never seen it above or below this coal bed. The name *filiculmis* is not well suited, as we find it with stems often anything but thread-like, they being half a cm. wide. Prof. Lesquereux has informed us by letter, that he intends to change the name in his forthcoming "Carboniferous Flora," which he is preparing for the Geol. Survey of Pennsylvania.

The occurrence of the plant at widely separated localities with the constant feature of depressed, shortened leaflets, precludes the idea that this is a consequence of any accidental distortion.

Sphenophyllum densifoliatum, Sp. nov., Pl. I, Fig. 7.

Stems, rather slender, containing numerous closely placed whorls of leaflets; whorls containing four leaflets; leaflets narrowly oblong-cuneate, united in pairs for a short space above the point of attachment, cut at the extremity into two short, closely approximated lobes, which have each two teeth; nerves single in the base of each leaflet, forking near the insertion, and each branch forking again a short distance above, and sending a long branch into each tooth at the end of the leaflet.

Habitat.—Roof shales of the Waynesburg Coal, Cassville, West Virginia.

Sphenophyllum tenuifolium, Sp. nov., Pl. 1, Fig. 9.

Stem slender, furnished with rather remotely placed whorls of six leaflets; leaflets, linear-cuneate, six-toothed at the end, without lobes; nerves, single in the base of the leaflet, forking three times above, so as to send a branch into each tooth of the leaflet.

This form is so well marked, and different from any species hitherto described, that we are compelled to assign it specific value. It occurs often by itself, with constant features; hence it cannot be an abnormal form of some other described species. It is a little like Germar's *S. angustifolius*. There is no connecting link to unite it with *S. densifoliatum*.

Habitat.—Roof shales of the Waynesburg Coal, Cassville and West Union, West Virginia.

Sphenophyllum longifolium, Germar.

This species is not rare in the roof shales of the Waynesburg Coal, at Cassville, and West Union. Some of its leaves have been seen which were more than an inch in length.

Sphenophyllum oblongifolium, Germar.

This is another species which is not uncommon. It is seen in great quantities in the shale parting which supports the roof coal of the Waynesburg bed at Cassville, West Virginia. Though often associated with *S. filiculmis*, it always presents an entirely different aspect, and they are without doubt different species.

ANNULARIA, Sternb.

This genus is well represented in the Upper Carboniferous strata, ascending into the highest beds, where *Annularia longifolia*, Brongt, is one of the most common plants. The following species have been seen:

Annularia carinata, Guth.

This is a very abundant form, and has been seen at Cassville, and at West Union, in the roof-shales of the Waynesburg Coal, and at Bellton, 400 feet above this coal bed. From West Union we have some very fine specimens of the plant, showing a stem bearing many long leafy branches. It is exactly like Gutbier's species.

Annularia longifolia, Brongt.

This species, as previously stated, ranges throughout the entire thickness of the strata above the Waynesburg Coal. It becomes more abundant towards the top of the series, where so many other plants, common at lower horizons, have disappeared.

Habitat.—Roof-shales of the Waynesburg Coal, at Cassville, West Union, and throughout the Upper Barren Measures.

Annularia sphenophylloides, Ung.

This well marked species has been seen at only one locality and horizon, and that was in the roof shales of the Waynesburg Coal, at Cassville, West Virginia.

Annularia radiata, Brongt.

At Cassville, West Virginia, in the roof-shales of the Waynesburg Coal, we find a very delicate species of annularia which very much resembles A. radiata Brongt. It is smaller, and the leaves are narrower, but the difference is not sufficient to separate the two.

Annularia minuta, Brongt.

This well marked little species was seen in immense quantities in the roof shales of the Washington Coal, near Little Washington, Penna. The leaflets are very short, and the joints of the stem seem to be swollen at the points of attachment. Though not seen in West Virginia, the nearness of the locality in Pennsylvania to the West Virginia localities, entitles it to mention here as throwing light upon the flora of our Upper Barren Measures.

FERNS.

SPHENOPTERIS, Brongt.

The Sphenopterids are by no means so fully represented in the flora of the Upper Carboniferous, either in the number of species, or individuals, as are the Pecopterids. The specimens of this genus are found sparingly, and the individuals are never, as is the case with many Pecopterids, abundant enough to form the preponderating element in the flora of any locality. The facies has changed greatly from that shown in these ferns at lower horizons. We find no species retaining the composite type shown in the Sphenopterids of the Conglomerate and Lower Coal Groups, as exemplified in S. macilenta, S. latifolia, &c. The facies of the plants of the upper beds seems rather to belong to horizons higher even than the Carboniferous, and reaching into the Rhaetic, and Oolite. It is a peculiar fact that we find, as yet, no well characterized Sphenopterid above the horizon of the Waynesburg Coal, although in many localities the shale is of a nature fitted to preserve the most delicate plants.

Sphenopteris acrocarpa, Sp. nov. Pl. III, Figs. 1–3; Pl. IV, Figs. 1–5.

(Frond, tripinnate; primary pinnae, triangular, or lanceolate in outline, curving upward from the rachis at an acute angle; secondary pinnae, sub-alternate, long, narrow, and somewhat pointed, the lowest one on the upper side being the longest and most complex in division, and extending up parallel with the principal rachis; pinnules of the lower and middle portions, lanceolate in outline, acute, and laciniate on the margin, the incisions making a very acute angle with the mid-rib of the pinnule, contracted at the base, and attached under a very acute angle to a narrowly winged rachis; laciniae of the lower pinnules of the frond, and pinnae, notched and toothed; of the middle portion, passing into teeth; and in the upper pinnae, being lost, causing the incised pinnules, to pass into small ovate ones, with entire margins; mid-nerve of the pinnule, somewhat flexu-

ous; lateral nerves, passing off at an acute angle into the
segments or laciniæ, pinnately divided, or forking; fructification, placed on the terminal lobe of the pinnules, at the
extremity of the median nerve, and consisting of six sori,
grouped radially around a central axis.)

The primary, and the secondary rachis, are both beautifully channeled on the upper side, and this is a feature so
constant, that we may recognize fragments by its means
with certainty. After the figures of the plant given on
Plates III and IV had been engraved, fertile pinnæ were
found, showing the character of the fructification much
more clearly, than that given on Pl. III. We are fortunate in possessing a large number of specimens of this
plant, as its complex character, and the great changes that
it exhibits in passing from the lower to the upper part of
the frond, would lead otherwise to the foundation of several species upon the different parts of this single plant.
Indeed it is difficult to do justice to it, either in a short description, or without using many figures.

The star shaped arrangement of the sori, seems to ally
the plant with *Asterocarpus*, of Weiss, and the general facies and nervation, with *Sphenopteris denticulata*, Brongt,
from the Oolite Formation. The terminal position of the
sori causes it to resemble the Hymenophylloid section of
the Sphenopterids, and in this point, it reminds us of
Schenk's *Acropteris cuneata*, from the Rhaetic.

Habitat—Found only in the Roof shales of the Waynesburg Coal, at one coal mine, at Cassville, W. Va.

Sphenopteris coriacea, Sp. nov., Pl. V, Figs. 5 and 6.

(Frond, bipinnate; pinnæ, inserted at an acute angle on
the broad, leathery, winged primary rachis, and terminating
at the summit in a three-lobed leaflet; pinnules triangular in outline, somewhat contracted at base, and decurrent
on the winged secondary rachis, cut into 3 or 4 rounded
lobes, the uppermost one being somewhat elongated; lateral
nerves obscure, or wholly concealed in the thick leathery-
like parenchyma of the lobes.)

This very peculiar plant has, as yet, been found only in

the roof-shales of the Washington Coal, 175 feet above the Waynesburg Coal. It occurs in company with *Callipteris conferta*, Brongt, covering with its leathery pinnæ, the surface of a thin layer of calcareous iron ore. It seems to be allied to *Sphenopteris oxydata*, Goepp, and *Sphenopteris lyratifolia*, Weiss, both of which occur in the Permian of Europe. Unfortunately the maceration, to which the fragments have been subjected, disguises somewhat the details of the plant, and it is found only in fragments. In some features it resembles *Callipteris*, especially in its thick, dense parenchyma, in the immersion of the nerves, and in the occurrence of pinnules on the principal rachis.

Habitat.—Roof-Shales of the Washington Coal, near Brown's Bridge, Monongalia Co., W. Va.

Sphenopteris dentata, Sp. nov., Pl. V, Figs. 7–8.

Frond, bi or tripinnate; pinnæ, linear-lanceolate, alternate, going off at almost a right angle; pinnules, ovate, slightly contracted at the base, and furnished with sharply pointed teeth; primary nerve, faintly marked in some of the pinnules; lateral nerves, wanting, or concealed in the thick leather-like parenchyma of the pinnules.

This beautiful little plant, is closely allied to *Sphenopteris Sarana*, Weiss, but it differs from it in its more pointed pinnules, and its apparent want of lateral nerves.

Habitat.—Roof-Shales of the Waynesburg Coal, Cassville, W. Va.

Sphenopteris species? Pl. XI, Figs. 5–7.

The fragments depicted in Figs. 5, 6, 7, Pl. XI, may represent a new species, but they are too small and imperfect to fix the specific character. Fig. 6 resembles some forms of *Sphenopteris Lesquereuxii*, Newb, and may be identical with it, but the lobes of this plant are sharper, and more deeply cut, than those of the above named Sphenopteris.

Habitat.—Roof-Shales of the Waynesburg Coal, Cassville, W. Va.

Sphenopteris auriculata, Sp. nov., Pl. VII, Figs. 3–4.

Frond, bi or tripinnate; principal rachis, pretty stout,

and smooth; pinnæ long, linear-lanceolate, alternate, going off at almost a right angle, somewhat arched; pinnules, oblong-ovate, alternate, incisely-lobed, the lobes more or less dentate, the lowest lobe on the upper side being larger than the others and projecting along the secondary rachis imparts an auriculate character to the pinnule; mid-nerve, strong, and well marked; lateral nerves, branching dichotomously, a branch passing into each tooth of the lobes.

This plant seems more closely related to *Sphenopteris cristata*, Brongt. than any other described Sphenopteris, and like the latter, as remarked by Brongniart, it possesses characters which ally it with Pecopteris.

Some of the pinnules, like that shown in Fig. 3*b*, have no denticulations on the lobes, and then their resemblance to Pecopteris is more marked. These occur in the lower portion of the plant.

Habitat—Roof Shales of the Waynesburg Coal, Cassville, West Virginia.

Sphenopteris minuti-secta, Sp. nov., Pl. V, Figs. 1–4.

Frond, quadripinnate; secondary pinnæ, short, and triangular, going off at nearly a right angle from the stout primary rachis; tertiary pinnæ, oblong-linear; quaternary divisions (pinnules) small, alternate, narrowed at the base, and decurrent on the rachis, obliquely inserted and cut into very small, almost microscopic lobes, which in the lower pinnules are notched at the extremity, and in the upper ones entire and tooth-shaped; mid-nerves of the pinnules, rather stout at the base, and soon becoming attenuate. Lateral nerves, slender, passing into each lobe of the pinnule, forking in the lower lobes, and single in those toward the extremity of the pinnule.

The ultimate divisions of this plant are so fine that they can be followed only with the aid of a lens. The texture of the pinnules is thin and delicate. We were fortunate in finding it in shale of great fineness and evenness, so that it is most beautifully preserved, the impressions being as distinct as if engraved on stone. The plant differs widely from any Sphenopteris hitherto described in the extreme minuteness of its lobes.   This is the only Sphenopteris which comes up from a lower horizon into the upper beds, as it is found 20 feet below the Pittsburg coal near Wheeling, W. Va.   The resemblance in facies of the plant to the genus *Thyrsopteris*, Heer, from the Oolite, is very striking, and, so far as the form is concerned, it would belong to that genus.   The fructification however of Thyrsopteris is not found, and as the plants are so widely separated in time, it is best to place it among the Sphenopterids in the absence of proof of its Thyropteris character.

Habitat—Twenty feet below the Pittsburg Coal near Wheeling, and in the roof shales of the Waynesburg Coal, West Union, W. Va.

Sphenopteris foliosa, Sp. nov., Pl. V, Figs. 9–11.

Frond, tripinnate; secondary pinnæ, very long; linear-lanceolate, rigid; tertiary pinnæ, short, oblong lanceolate sub-opposite, inserted at an angle of 45°; pinnules, sub-quadrate or rotundate, decurrent, cut into slightly marked segments, which are notched into two rounded teeth, or are simple; mid-nerve, well defined but slender; lateral nerves passing off obliquely, and forking into the incisions, a branch passing into each tooth.

The plant has a thick, fleshy leaf substance, and belongs to the Pecopteroid section of the Sphenopterids, a section which seems most abundant in the upper beds.

Habitat.—Roof shales of the Waynesburg Coal, Cassville, West Virginia.

Sphenopteris Lescuriana, Sp. nov. Pl. VI, Fig. 1; VII, 1–2.

Frond, quadripinnate; rachis of primary pinnæ, stout and smooth; rachis of secondary pinnæ, strong, going off at right angles from the primary one, and arching slightly outwards; secondary pinnules, long, oblong-elliptical in outline; tertiary pinnæ, numerous, linear-lanceolate, alternate; quaternary pinnæ (pinnules) lanceolate, densely crowded, those near the base of the secondary pinnæ, again divided, with pinnatifid lobes or divisions, the lowest pinnule (quaternary pinna) heteromorphous, being larger and more complex in division than the rest, and deflexed along the secondary rachis, incisely lobed, with the divisions bluntly toothed, slightly decurrent on the tertiary rachis, and becoming united towards the end of the pinnæ; primary nerve of the pinnule, strong and somewhat flexuous, giving off nerves which branch palmately into the rounded, slightly marked lobes of the pinnules, a branch passing into each of the crenate teeth of these lobes.

The ultimate pinnae, or pinnules, at the lower part of the secondary pinnæ, are so much larger and deeply cut than the rest, that in them the plant is quinquepinnatifid at least. The tertiary pinnæ near the base of the secondary ones, are shorter than the normal ones, and have contracted pinnules, whose nervation is distorted somewhat, and shows a tendency to inflation, as if this portion of the plant might become fertile, but no fructification can be made out. The dwarfing of this portion of the pinnæ, is contrary to the rule, as we find generally that the length of the pinnæ diminish from the base to the summit of the rachis which bears them. Another of the curious features of this plant is the marked heteromorphism and deflexed position of the basal pinnules on the lower side of the pinnæ. The pinnules on the lower portion of the pinnæ are slightly decurrent, and united each by a narrow wing to the next lower, while towards the summit they are more and more united. The value of the peculiar dwarfing of the lower portion of the secondary pinnæ we cannot determine, as we found only one specimen showing this part of the frond. It may be specific. It will be observed that this portion is preceded by a pair of large complex pinnæ, such as we might expect from their position, which is next to the primary rachis.

This plant, which is one of the finest in the entire Carboniferous flora, is beautifully preserved in the fine grained shale, on which it is found, and every detail can be easily made out. Its affinities seem to be with *Pecopteris cristata*, Brongt., which it resembles in some points, and it evidently belongs to the Pecopteroid section of the Sphenopterids.

The existence of such well marked types, uniting the features of Sphenopteris and Pecopteris, as was first noticed by Brongniart, would seem to call for their separation into a sub-genus, which as the Sphenopteris facies is that best marked, might be styled *Sphenopteris-Pecopterides*.

The plant is named in honor of the eminent palæo-botanist, Prof. Leo Lesquereux.

Habitat — Roof shales of the Waynesburg Coal, West Union, W. Va.

Sphenopteris pachynervis. Sp. nov., Pl. VII, Figs. 5 6.

Frond, bi or tripinnate, pinnæ opposite, going off from the main rachis at an angle of about 50°, lanceolate in outline; pinnules are alternate, closely set, and incisely lobed, the lobes often toothed; primary nerve very thick; lateral nerves, very large, passing into each of the lobes of the pinnules, and usually forking once or twice.

This plant also resembles the type shown in *Pecopteris cristata*, Brongt., but is sharply distinguished by the great size of its nerves. The texture of the parenchyma is coriaceous, and this serves to exaggerate somewhat the nerves.

Habitat—Roof shales of the Waynesburg Coal, West Union, W. Va.

Sphenopteris hastata. Sp. nov., Pl. VII, Fig. 7.

Frond, bipinnate; pinnæ, long and linnear; rachis of pinnæ, slender and terete; pinnules, alternate, lanceolate, with a somewhat hastate base, formed by a sudden contraction at the insertion of the pinnules; lobes, on each lamina 4 to 5, with the lobes possessing two or three teeth; mid-nerve, rather strong; lateral nerves, rising at an acute angle into each lobe of the pinnule, and forking so as to send a branch into each tooth of the same.

Habitat — Roof shales of the Waynesburg Coal, Cassville, W. Va.

NEUROPTERIS, Brongt.

The only species of this genus which pass from the lower to the upper beds above the Pittsburg Coal, are those cosmopolitan forms N. hirsuta and N. flexuosa. These range nearly through the entire carboniferous formation. They extend to the highest portions of the Upper Barren Measures where plants are found, forming by far the larger part and sometimes nearly all of the flora of the highest horizons.

Neuropteris hirsuta, Lesqx., Figs. 7 and 8, Pl. VIII.

This species is one of the most abundant plants at all horizons and at all localities in the Upper Barrens. It is a noteworthy fact, that where it and N. flexuosa abound, we rarely find many other species, these two plants seeming to exclude the small foliage ferns, such as *Sphenopteris*, *Pecopteris*, &c. This peculiarity is so marked that, while in the particular part of the stratum holding the neuropterids in question we find no other ferns, yet in a layer above or below, deposited under different conditions, and separated by but a few inches of space from the first named, we often find great numbers of Pecopterids &c. but no Neuropteris hirsuta and N. flexuosa. Their mode of growth and exposure to transport by water must have been totally different from those of most other genera of ferns.

The hirsute character of *N. hirsuta* is almost never seen in the upper beds; but the form, nervation &c. are identical with those of the plant which shows this feature when found at lower horizons.

During our researches into the flora of the various beds of this and adjoining States we were so fortunate as to find undoubted fruiting forms of *N. hirsuta*, and are thus able to throw some light on the character of the fructification of this important genus, a point which has long remained in obscurity.

This fructification, as shown in Figs. 7 and 8, Pl. VIII, consists of linear-elliptical sori, 4½ mms. long and 1 mm. wide at the middle. They are normally placed in groups of four, the sorus nearest the base of the pinnule being situated near the middle of the lamina of the pinnule, while each succeeding sorus of the group approaches nearer to the mid-rib, until the last one comes quite close to it, thus forming rows, each containing 4 sori, and each row inclined towards the mid-rib. The general method of fructification is very similar to that of *Scolopendrium vulgare*, the sori appearing to lie between adjoining branches of adjacent nerves. They have a raised margin on each side, which closely resembles the double indusium of Scolopendrium.

Bunbury, in vol. III. Quar. Jour. Geol. Soc. on Pl. 21, figures a hirsute plant from Cape Breton, which he considers as Neuropteris cordata. It is plainly identical with N. hirsuta. On it are depicted depressions like those on our plant, though smaller, and showing a similar arrangement, i. e. groups of four (when complete), the uppermost depression being nearest to the mid-rib. He states that they lie between the veins, and thinks them the result of disease. They are probably impressions of sori as in our plant.

Brongniart, in his Hist. d. Veg. Fos. Pl. LXV, Fig. 3, gives what he considered as the fructification of N. flexuosa. The arrangement of these markings seems to be without definite order, and judging from their general character, they appear not to form fructifications. They do not agree with the fructification given for this plant by Dr. Heer in his "Uhr. d. Schweitz ; " Die Pfl. d. Steink. Periode." We have seen a pinna of N. flexuosa, containing 6 pinnules, each pinnule marked by a row of elliptical elevations on each side of the mid-nerve. These, which are evidently impressions of sori, agree essentially with Heer's fructification, but they are larger and more elongate elliptical in shape. This specimen was in the collection of Mr. Gustav Gutenberg of Wheeling, and was collected at the locality affording the fructified N. hirsuta. Mr. Gutenberg kindly offered to place it in our hands for description and figuring, but it was unfortunately lost before reaching us.

The specimens figured in Figs. 7 and 8, as well as others of the same character, were found by us near Bellaire, Ohio, 20 feet below the Pittsburg Coal, in a very fine grained shale which has beautifully preserved the plants contained in it.

Neuropteris flexuosa, Brongt.

This plant is one of the most widely diffused and persistent of all the Carboniferous flora. It ascends from the Vespertine, where, (as at Lewis Tunnel,) it is slightly modified, to the top of the Carboniferous system. In the upper beds, it forms by far the most abundant plant, often excluding all others from certain localities.

Plate VIII. Fig. 6, shows a very singular form of this species, which would, if found isolated, be taken for a different species; but so many intermediate forms, connecting it with the normal plant, occur, that it cannot be separated from it. The pinnules are very small, somewhat falcate, and attached by all of the somewhat contracted and rounded base. This form is found in the roof-shales of the Waynesburg Coal, at Carmichaels, Penna.

Plate VIII. Fig. 1, shows a form of flexuosa, which differs from the normal type sufficiently to constitute a variety at least. This may be styled: *Neuropteris flexuosa*, var. *longifolia*. It is distinguished from the typical forms of the species, by having much longer pinnules, which are also *opposite*, a feature not seen in N. flexuosa, or indeed commonly in Neuropterids. It is not a new species, for we find intermediate forms connecting this with the normal type.

Habitat.—Roof-shales of the Waynesburg Coal, West Union, W. Va., with great numbers of the normal form.

Neuropteris dictyopteroides, Sp. nov., Pl. VIII, Figs. 3–5.

Frond, pinnate or bipinnate; pinnules, alternate lanceolate, with cordate base, and attached by a cordate base to a rather stout rachis; mid-nerve, very broad and appearing to be made up of parallel adjoining nerves, which are formed by the union of the lateral nerves, producing a flat ribbon-like bundle; lateral nerves, dichotomizing in passing to the margin as usual in Neuropteris, very fine, and rather indistinct in their course, sending off delicate thread-like branches which anastomose with the adjoining lateral nerves, at an acute angle and forming elongate meshes.

4 PP.

The delicate thread-like branches, on leaving the lateral nerve, rise somewhat towards the surface of the parenchyma of the leaf, and unite with the adjoining lateral nerve, on the upper side of it, thus giving the nervation of the plant a peculiar aspect; for we can detect the Neuropteris nervation under what appears to be a net work of delicate thread-like branches, which partly overlies it. At first sight these delicate branches might be taken for hairs, but they are plainly off-shoots from the lateral nerves. The *plant* has thus the appearance of a Dictyopteris. Von Röhl figures in his "Fos. Fl. d. Steink. West." &c., Pl. XV, Fig. 6, Pl. XXI, Fig. 7*b*, a plant which would seem to be close to ours, and which he calls *Dictyopteris neuropteroides*.

Habitat.—Roof shales of the Waynesburg Coal, West Union; Bellton, Marshall county, 400 feet above the Waynesburg Coal.

Neuropteris auriculata, Brongt.

This species seems quite widely distributed in the Upper Carboniferous strata, though seldom found in great abundance. At Cassville, in the roof shales of the Waynesburg Coal, we find a fern which agrees quite closely with Brongniart's Neuropteris Villersii, which, as Schimper correctly states, is identical with auriculata, for our plant passes into the typical auriculata. Neuropteris auriculata is abundant at West Union in the roof shales of the Waynesburg Coal, and passes up high in the Upper Barren Measures, occurring at Bellton 400 feet above the Waynesburg Coal, and at other localities.

Neuropteris odontopteroides, Sp. nov., Pl. IX, Figs. 1-6.

(Frond, pinnate; rachis, very stout, and broad; pinnæ, alternate, or sub-opposite, oblong-ovate, or lanceolate, going off nearly at a right angle, attached by the lower part of the base, the upper being free and slightly cut away, which, with the cutting away of the end of the pinnule on the lower side, gives a squamose aspect to the same;

toward the summit of the pinna, the broad rachis widens
out into a heteromorphous terminal pinnule, which is
usually somewhat falcate, and slightly eared at base, or
lobed occasionally, by consolidation with the adjoining
small pinnules of the terminal part of the pinna; the
upper pinnules tend to unite and pass into lobes, by growing smaller and being attached by their whole base; mid-
nerves distinct to about one-quarter the length of the pin-
ule, and formed of three principal nerves consolidated together; these three principal nerves split up and repeat-
edly branch, until they fill the entire pinnule; at their
lower portion, by consolidation, they form a broad, flat,
strap-shaped bundle of nerves, which in their insertion
occupy a considerable portion of the base of the pinnule;
in the upper pinnules, the nerves go off from the entire base,
as in Odontopteris; they are tolerably strong, and are very
distinct.)

This plant has a facies much like *Odontopteris Dufresnoyii* (Brongt.) Schimp., in nervation, and modes of
attachment of the pinnules. From the great size of the
rachis and other points it would seem to be simply pinnate,
and apparently to belong to Schimper's sub-genus *Neuropteridium*, which contains plants belonging to the Trias.

Habitat—Roof-shales of the Waynesburg Coal, Cassville
and West Union, W. Va.

Neuropteris fimbriata, Lesqx.

In the roof-shales of the Waynesburg Coal, at Carmichael's, Penn., we find very fine specimens of this plant
2 inches long and as many wide. It occurs there associated
with Neuropteris flexuosa.

Neuropteris species? Pl. X. Fig. 11.

This huge leaflet was found associated with abundant remains of N. flexuosa near Jolleytown, Greene Co. Penn.
400 feet above the Waynesburg Coal. It is probably a ra-
chial leaflet of N. flexuosa.

Neuropteris cordata, Brongt.

This plant, as limited by Schimper, is quite common in

the roof shales of the Waynesburg Coal at West Union,
forming with N. dictyopteroides, the greater part of the
plants found in some layers. It is very abundant at Cassville at the same horizon. The pinnules are often of very
large size reaching the length of 6 inches and the width
of 1½ inches. It has often a falcate form, and reminds us
of the shape of N. Rogersi of Lesqx., but is proportionally
longer. It is confined to horizons above the Pittsburg Coal
in West Virginia.

ODONTOPTERIS, Brongt.

This genus is represented in the Upper Carboniferous
flora of W. Va. by only a few species, mostly of types
found in the Permian of Europe.

The individuals, except of one species, are also rare in
occurrence.

Odontopteris obtusiloba, Var. *rarinervis*. Pl. X, Fig. 4.

The fragment, depicted in the above named figure, differs
from *Odontopteris obtusa*, Naum, as figured by Weiss,
Geinitz, and others, only in having fewer nerves, and a
stronger rachis. The overlapping of the pinnules is precisely the same with that shown by Geinitz: "Dyas," Pl.
XXVIII, Figs. 1—4. It occurs in a bed of red shale 400
feet above the Waynesburg Coal, at Bellton, Marshall Co.,
Va.

Odontopteris nervosa, Sp. nov. Pl. X, Figs. 1–2.

(Frond bi-pinnatifid; pinnules very deciduous, oblong-
linear, at the insertion cut into rounded or oval segments
nearly to the rachis, the segments towards the extremity
less and less deeply cut, at the extremity entire; nerves
rather distant, very strong, and sharply defined, passing
from the entire base of the lobes with no median nerve.)

The plant has a striking similarity to some forms of the
Odontopteris obtusa, but is distinguished by the fewer and
coarser nerves. Fig. 2. shows the tendency of the terminations of the pinnules to become entire, a feature shown
also in *O. obtusa*. The specimens found, which are not un-

common at West Union, are almost always fragments of pinnules as in Fig. 8. They have never been seen attached except in a single specimen (Fig. 1) seen at Cassville.

Habitat—Roof Shales of the Waynesburg Coal, Cassville, and West Union, W. Va.

Odontopteris pachyderma, Sp. nov., Pl. X. Figs. 5-10.

(Frond, bipinnate; primary rachis, stout, secondary rachis, slender and delicate; pinnæ, alternate and somewhat closely placed, very deciduous, going off at nearly a right angle; pinnules, oblong and ovate, inclined forward, sometimes falcate, the lowest one on the lower side heteromorphous, being bilobed, apparently formed of two consolidated pinnules, the lowest ones on the upper side occasionally heteromorphous, the lowest pinnules of the lower part of the plant separate nearly or quite to the base, with rounded lobes or undulate on the margins, pinnule of the middle and upper portions united at base, ovate and acute, becoming more united towards the summit of the frond where the pinnæ pass into pinnules, and also toward the end of the pinnæ, being almost entirely united at the extremity; leaf-substance, exceedingly thick and dense; midnerve more or less distinct and splitting up dichotomously into branches which diverge in an angular manner; lateral nerves coming off from the principal rachis also, and branching dichotomously, all very delicate, and almost always concealed in the dense parenchyma.)

The nerves are so fine, and the leaf substance so dense, that out of the large number of specimens examined only one or two showed the details of the nervation.

The plant usually leaves a dense shining film on the shale. The form of some of the pinnæ and pinnules strikingly resembles Heer's *Pecopteris triassica*, "Pfl. d. Trias u. des Jura," Pl. XXV. Figs. 1 and 2, but the nervation is totally different. The singular nervation approaches nearer to that of Odontopteris alpina, Heer. It would come in the subgenus, O. Mixoneura of Weiss, as would all the species found as yet in the upper beds of West Virginia.

Habitat—Abundant in the roof shales of the Waynesburg Coal at Cassville, W. Va., also found abundantly 500 feet above this horizon in Greene Co. Pennsylvania.

Odontopteris densifolia, Sp. nov., Pl. X, Fig. 3.

Frond, pinnate, or bipinnate ; pinnules, ovate, inclined forward, densely placed, touching by their borders so as to appear imbricated, nerves, going off from the entire base ; at the center a bundle of nerves issues, which is quickly dissolved into branches, all exceedingly fine but distinct, dichotomosing again and again so as to fill in a flabellate manner the end of the pinnules.

The issuing of the bundle of nerves at the middle of the base of the pinnule gives the appearance of a short midnerve.

The leaf substance is very dense and thick.

Habitat—Roof shales of the Waynesburg Coal, Cassville, West Virginia.

CALLIPTERIS. Brongt.

Callipteris conferta, (Sternb.,) Brongt. Pl. XI. Figs. 1–4.

The plant which we have identified with *Callipteris conferta*, is found in considerable quantities, covering the surface of a calcareous iron ore which occurs in the roof of the Washington Coal, 175 feet above the base of the Upper Barren Measures, near Brown's Bridge, Monongalia Co., W. Va. It is associated with *Sphenopteris coriacea*, F. & W., and these two plants form almost the entire flora at the place mentioned. We find there all the forms of *C. conferta* that have been figured by Weiss in his "Foss. Fl. d. Jünst. Steink. u. Roth." except the fruiting form which he gives. The plant, as we find it, is very thick and leatherlike. There seems to have been a fleshy epidermis extending over the rachis in the upper portions of the plant, which caused this to appear much wider than it really is, as is shown in Fig. 4*a*, which represents an enlarged portion of Fig. 4. The lower part gives the appearance of the pinna when the thick epidermis is removed, and the upper part when it is present. It seems to be caused by the extension of the epidermis of the leaflets over the rachis. The nerves are immersed in the dense leaf-substance, which may be pulled off from the stone in flakes, leaving an imprint in which no sign of the lateral nerves appears. The nerves of the broader and larger pinnules appear to fork once, while those of the narrower and longer pinnules, which appear in the middle portions of the frond, seem simple. They are usually so enclosed in the dense leaf-substance as to betray their presence by creases, which no doubt are much stronger than the true nerves. As we could see only these creases, indicating the course of the nerves, we have depicted them in the lateral nerves given in the figures. The nervation as thus determined, appears to agree with that given by Weiss. We cannot agree with Weiss in referring the plant to *Alethopteris*, as it is entirely different from any species of that genus known to us.

This is the only species of true Callipteris that has ever been found in the Appalachian Coal Field, and its appearance marks an important change in the flora of the horizon containing it. The other species from this field attributed to Callipteris are either Alethopterids, like A. Sullivanti, or plants of the Pecopteris type, and belonging to *Callipteridium*.

In Europe this plant is regarded as a characteristic Permian species.

CALLIPTERIDIUM, Weiss.

Dr. Weiss has established under the name of Callipteridium a genus to include plants which, with the facies of Pecopteris, have a nervation resembling that of Neuropteris. Schimper gives the following as the generic character: "Median-nerve of the pinnules, strong, vanishing; secondary nerves obliquely spreading, simply or doubly forked, parallel to each other."

These plants are amongst the finest found in the Upper Carboniferous strata, and in W. Va. are peculiar to them,

for no Callipteridium has been found in this State below
the Pittsburg Coal.   Many of the Pecopterids of the Upper
Carboniferous show a tendency to assume the characters
of Callipteridium.

Callipteridium Dawsonianum, Sp. nov., Pl. XIII, Figs.
1–2, and Pl. XIV, Fig. 1.

(Frond, tripinnate, or tripinnatifid; rachis of the primary
pinna, strong and rough ; secondary pinnæ, lanceolate-
linear, alternate, going off at an angle of from 45°—60° ;
secondary rachis, rather strong : pinnules, united at the
base, ovate and inclined slightly towards the apex of the
pinna, terminal pinnule, rounded-elliptical, or obovate,
lowest pinnule on the lower side, half inserted on the primary rachis ; mid-nerves of the pinnules, strong, and vanishing toward the apex of the pinnule ; lateral nerves,
forked near the insertion, and again forking near the middle of the lamina, arising at a very acute angle, the branches
curving out and passing off nearly parallel to each other,
the lowest nerves of adjacent pinnules meeting at the
sinus, several nerves arising from the secondary rachis below the insertion of the mid-nerve.

This splendid plant, which we have named in honor of
Principal J. W. Dawson, the distinguished paleo-botanist,
is very large, and must have been arborescent.   We find
its huge fronds spreading over the surface of the shale, and
the fragments seen are sometimes 2 feet wide and 3 feet
long, as was the case with the specimen of which one of
the primary pinnæ is depicted in Fig. 1, Plate XIII.   The
texture of the pinnules seems to have been coriaceous, and
thick.   The nervation and facies of the plant are much like
those of Heer's *Meraniopteris angusta*, figured in his "Pfl.
d. Trias u. Jura," Pl. XXXVII.

Habitat—Roof shales of the Waynesburg Coal, West
Union, West Virginia.

Callipteridium oblongifolium, Sp. nov., Pl. XII, Figs.
1–5.

(Frond, tripinnate : secondary pinnæ rigid, and rather

remote ; rachis of the primary pinna, large and very rigid,
that of the secondary, rather slender; pinnules of the
lowest portion of the frond, undulate on the margin as
if about to become lobed, those of the middle and upper portions very entire, all oblong, obtuse, and very rigid
with a dense leather-like leaf-substance, slightly cut away
at the base on the upper side, inserted at an acute angle on
the secondary rachis, alternate ; mid-nerve, strong at the
lower portion, but splitting up and vanishing toward the
end ; lateral nerves in the lowest pinnules, grouped flabellately in the lobes or forking with parallel branches, and
the two adjoining nerves inserted at a common point, those
of the pinnules higher on the frond forking near the insertion and sending parallel branches to the margin ; fructification consisting of two rows of sori, one on each side of
the mid-nerve, elliptical in form, and leaving very sharply
defined pits on the surface of the pinnules ; fertile pinnules
thick and dense, with no lateral nerves, and an obscurely
defined mid-nerve.)

The form, and sharp definition of the impressions of the
sori, with the nervation and shape of the sterile pinnules,
cause our plant to resemble Heer's Asterocarpus (Pecopteris) Meriani, Pl. XXIV. Figs. 4. 5, 6, " Die Pfl. d. Trias
u. des Jura," but in our species the pinnules are separate
to the base.

The fertile form given in Pl. XII, Fig. 2, we attribute to
C. oblongifolium ; though we have never seen it attached to
the sterile portion of the plant. we find it with the sterile
portions of the plant, and the resemblance in facies is evident. We find the fructified part of the plant, with the
leaf-substance preserved on the stone. This seems to have
been thick and leather-like. for the organic matter now re-
maining presents the form of a dense shining film, in which
we find the elliptical pits showing the places of the sori.
These pits are beautifully distributed, and seem to indicate
that the sori were placed in indentations on the surface of
the lamina. As the sori themselves seem to have fallen
out before the entombment of the plant we could not detect their nature.

Habitat—Roof-shales of the Waynesburg Coal, Cassville, W. Va., and at Bellton, 400 feet above the Waynesburg Coal.

Callipteridium grandifolium. Sp. nov., Pl. XV, Figs. 1–4, and Pl. XVI, Figs. 2–4.

(Frond, tripinnate; primary rachis, strong and rough; secondary pinnæ, going off at an acute angle, alternate, approximate, linear-lanceolate in outline; pinnules, closely placed, quite variable in shape, but normally oblong, often passing into elliptical forms more or less broad, slightly narrowed by being cut away on the upper side of the base, and rounded on the lower side, obtusely rounded at the end, separate except towards the summit of the frond and pinnæ, alternate; those of the lower part of the plant, slightly round-lobed, the lobing irregular in the number of lobes on different sides of the same pinnules and on adjacent pinnules; terminal lobes of the pinnæ, round-elliptical and united with the adjacent pinnules; pinnules towards the summit of the frond becoming more united and smaller, so that the ultimate pinnæ grow shorter and less deeply lobed and finally pass into pinnules of the normal kind; lower pinnule on the lower side of the pinnæ usually inserted half on the secondary rachis; lower pinnules on the upper side of the ultimate pinnæ often somewhat heteromorphous; (Figs. 3 and 4), mid-nerve, strong towards the base and splitting up towards the end; lateral nerves, rising at a very acute angle, forking near the insertion and again about the middle of the lamina, arching off suddenly and strongly and passing with the branches nearly parallel so as to meet the margin at nearly a right angle, several nerves passing from the secondary rachis; fructification composed of elongate sori, placed on or between the branches of the lateral nerves, and extending nearly from the mid-nerve to the margin of the pinnule.)

This plant, from the great size and width of its pinnules, its numerous nerves, and vanishing mid-nerves, at first sight might be taken for a Neuropteris, but it clearly belongs to the genus Callipteridium. From the size, and arrangement of the parts of the plant, as seen *in situ*, it must have been one of the largest of the Carboniferous ferns, surpassing in size C. Dawsonianum. Owing to the nature of the containing shale it could only be obtained in a rather fragmentary condition. The leaf-substance was exceedingly thick and leather-like, leaving deep indentations in the shale. The fructified pinnules were too poorly preserved to show the details with distinctness. The rounded lobes of the lowest pinnules, owing to the thick nature of the leaf-substance, and pressure into the yielding mud on which they fell, often have their margins curved down, causing the surface of the lobe to stand out in relief. Fig. 1 shows the lowest part of the frond, with the irregular lobing; Figs. 2 and 3 show portions of the middle of the frond; Fig. 4, shows a somewhat higher portion, and Fig. 2, Pl. XVI, gives the summit of the primary pinnæ. Fig. 4 shows that the fertile pinnules are more distinct, and more contracted at base, than the normal sterile pinnules.

This plant in several features has a close resemblance with Geinitz's plant which he figures in Steinkoh. von Sachs. Pl. XXXII, Figs. 1–5, as Alethopteris (Pecopteris) pteroides, and which as Grand 'Eury correctly says, seems to be a different plant from Brongniart's P. pteroides. The points of difference however are too numerous to permit us to unite them.

The fructified pinnæ, and the mode of fructification, are most strikingly like those of *Asplenites Ottonis*, Schenk, from the Rhaetic, and given in his "Foss. Flor. d. Grenzsch," &c., Pl. XI, Figs. 1 and 2. It forms another of the many plants which we find in the Upper Carboniferous of West Virginia, foreshadowing in a striking manner Triassic and Rhætic types.

Habitat.—Roof shales of the Waynesburg Coal, West Union, West Virginia.

Callipteridium odontopteroides, sp. nov.. Pl. XVI, Fig. 1.

(Frond, bi or tripinnate; principal rachis, slender; primary pinnæ, or frond, elliptical in outline; pinnæ, numerous, crowded together and growing shorter towards the

summit, and passing into simple pinnules, linear; secondary rachis, slender; pinnules, united at the base, oval, inclined forward, or falcate, becoming more united towards
the end of the pinnæ, and towards the summit of the frond
or pinna, and smaller, until near the apex the ultimate
pinnæ have passed into pinnules of the normal kind; lowest pinnule on the lower side, heteromorphous, and approaching the character of Odontopteris, being also inserted partly on the primary rachis, and toward the summit of the frond having no mid-rib; mid-nerve, well
marked, and splitting up towards the end; lateral nerves,
numerous and going off from the mid-nerve at a very
acute angle, forking once dichotomously, the lowest branch
of the lowest nerve, on the upper side of the pinnule, turn-
ing up into the sinus of the united pinnules; lateral nerves,
passing off from the principal rachis, one or more.)

This singular plant combines with the type of Callip-
teridum some marked features belonging to Odontopteris,
as the great length of the pinnæ, their peculiar method of
passing into pinnules near the summit of the frond, and
the heteromorphous lower pinnule. It is quite distinct
in facies from any other plant in the upper strata. The
texture of the pinnules seems to have been thin and delicate, and the nerves, though slender, to have been sharply
defined. It was evidently a large arborescent plant, as the
fragment figured was only a primary pinna.

Habitat.—Shales some 15 feet above the Waynesburg
Coal, near Arnettsville, West Virginia.

Callipteridium unitum, sp. nov., Pl. XIV, Figs. 2 and 3.

(Frond, tripinnate, or tripinnatifid; pinnæ, going off
acutely, somewhat deflexed; primary rachis, stout and
rigid; secondary rachis, rather slender; pinnules, near the
base of the pinnæ, and especially on the lower side, cor-
date-ovate, from the laminæ being constricted above the insertion; the rest, falcate, inclined forwards, all united, the
union increasing towards the ends of the pinnæ; lowest
pinnule, on the lower side, usually deflexed along the primary rachis; mid-nerve distinct, but soon dissolving into

branches; lateral nerves, near the base twice-forked, in ascending only once forked, with one of the branches again forking, uppermost nerve only once forked.)

This plant is peculiar in many respects. The nerves, and facies of the pinnules, resemble the genus *Cladophlebum* established by Schimper for certain peculiar Pecopterid forms which characterize the Rhætic and Oolite formations. Its heteromorphous pinnules ally it to Odontopteris. It is much like the plant described by Weiss in his Foss. Flor. d. Stein. und Rothl. under the name of *Neuropteris cordato-ovata*, Pl. 1, Fig. 1.

Our plant evidently belongs to the same type as Weiss'. Neither of them are truly neuropterid, and indeed there is no genus which will properly include this composite type. It would seem best to form a new genus to receive them, which could thus include all plants with the pinnules of Pecopterid type, combined with heteromorphism in the lower ones, and containing the nervation of Neuropteris. This might be styled "*Pecopteridium*." The generic character would then be: Frond bi or tripinnate; pinnules united at base, the lower pinnules on each pinnæ constricted near the base, the rest ovate and attached by a broad base; midnerve quickly dissolved into branches; lateral lower nerves twice forked, upper nerves once forked.

We place it however provisionally in the genus Callipteridium.

PECOPTERIS. Brongt.

This genus, in the Upper Carboniferous Flora of West Virginia, is richer than any other in the number of species, and, with the exception of Neuropteris, in the number of individuals also. The section Cyatheides furnishes the greatest number of species and individuals. While some species occurring at lower horizons are found here, yet the facies, as a whole, is changed by the addition of many new forms, and we find ourselves compelled to add considerably to the already long list of Pecopterids.

There is a tendency to pass into the form of Callipteridium, even in cases where the departure is not sufficient to separate the plant from Pecopteris.   The pinnæ of the last order often assume an elongate, linear form, and the pinnules, a falcate shape not usually seen in the Pecopterids of lower horizons.   We also find that the lowest pinnule, on the lower side of the ultimate pinnæ, is often heteromorphous, and inserted partly on the rachis of superior order.   The forms occurring in older strata which do pass up into the upper beds are generally considerably changed, so as to present a different facies, though they retain the leading features on which their specific value depends.   In the case of some the change is so great that we may have erred on the side of conservatism in identifying them with species already described.

The plants of this genus which, being found in older strata, also occur at the higher horizons are those which immediately follow.

Pecopteris arborescens, (Schloth.) Brongt.

The representatives of this species are among the most abundant of the plants, occurring in the upper horizons. Among the many specimens seen, we find none that agree entirely with the typical form of the plant as seen in the lower horizons of the Carboniferous Strata.

Nearly all the forms seen are more closely allied to the Permian plant which Goeppert has described under the name Cyatheites Schlotheimii.   Many specimens are much more delicate and finely cut than Goeppert's plant.   The characters of this form are so constant and distinct that it may well be questioned whether it should not remain a distinct species.   We find this plant in every portion of the upper beds where fossils are found, as at Carmichaels in Pennsylvania, and at Cassville, West Union, Bellton, &c. in West Virginia, both associated with the Waynesburg Coal and at higher levels.   Along North Ten Mile Creek in Washington Co. Penn. it is very abundant at the horizon of the Washington coal.

Pecopteris arborescens, Var. *integripinna*, Pl. XXVII. Fig. 6.

This curious looking plant has its pinnules somewhat similar in form with P. arborescens, but the general facies is quite different. We at first supposed that it might have derived its peculiar character from some malformation, or some effect produced by compression and distortion, but we find it at three widely distant localities, viz: Tyler, Marshall and Monongalia counties, unfortunately always failing to show the minute details of its nervation, a point which if known would decide its specific character. It is not a fructified pinna of some large Pecopteris, for some of the specimens show that the seeming pinnules are really composed of united pinnules of the type of P. arborescens. As may be seen, the pinnæ are very short and broad, simulating pinnules in form. They show no tendency to lengthen as we pass to a lower part of the common rachis. The plant in Tyler and Marshall counties occurs from 400 to 600 feet above the Waynesburg Coal, and in Monongalia county it is found on the horizon of this bed.

Pecopteris Candolleana. Brongt. Plate XX, Figs. 1, 2, and 3.

West Union in Doddridge Co. is the only locality at which this plant has been seen. It occurs there in great abundance in the roof shales of the Waynesburg Coal. The facies of the plant differs somewhat from Brongniart's species, but closely resembles the plant figured by Germar. With these well known forms we find some which present differences sufficient to call for description. The ultimate pinnæ of what appears to be the lower part of pinnæ of a superior order, are seen to possess crenulated pinnules, and these more rarely pass into lobed pinnules, while the nervation becomes correspondingly more complex. The normal form has in the pinnules lateral nerves, forking once, or at most with one of the veinlets forking again, as Germar well shows. In the crenulated pinnules of our specimens the lateral nerves are twice forked. The lobed pinnules do not show the details of the nervation, owing to poor preservation.

The crenulated pinnules are quite long, being usually 2 centimetres in length. Fig. 2 shows the pinnules with crenulate margins, and Fig. 3, those with more pronounced lobes. The lower basal pinnules of the normal forms show a tendency to heteromorphism, in being elongated and depressed along the rachis to which its pinnæ is attached. We would call attention to the difference in the facies of Brongniart's plant from that of Germar. Our specimens show all the forms figured by Germar, almost in *fac-simile.*

It is worthy of note that we find the same form of fruiting pinnules with those given by Germar, only the impressions of the sori are larger. They seemed to be formed by inflations of the ends of the lateral nerves, and often occupy the entire space between these nerves. Fig. 1 represents a fruiting pinna, and 1*a* an enlarged portion of the same. The crenulate and lobed forms are very rare.

Pecopteris elliptica, Bunb. Pl. XVII, Fig. 1.

Several specimens of this rare and well characterized species were found in the roof-shales of the Waynesburg Coal at Cassville, Monongalia Co., but they are very rare here, and we have found the plant nowhere else. Bunbury found at Frostburg, Maryland, a few fragments of a plant which seems identical with ours. He gives figures of it in "The Quar. Jour. of the Geol. Soc.," Vol. II, 1845, under the name Pecopteris elliptica. Our specimens are much larger, and show the details and facies better, as figured on Plate XVII, Fig. 1 and 1*a*. Schimper is in error when he states that "the lateral nerves of the pinnules diverge strongly after forking," for in Bunbury's plant, as in ours, the divergence is quite slow. The plant seems to have had a very robust, rigid aspect, and thick leathery pinnules.

The strata for a considerable distance above the Pittsburg coal are exposed at Frostburg, and it is quite probable that the plant occurs there at the same horizon as at Cassville.

Pecopteris Oreopteridia (*Schloth.*), Brongt.

In the Upper Barren Shales, at Bellton in Marshall Co.,

400 feet above the Waynesburg coal, we find a form of Pecopteris which we identify with the above species. It has precisely the character of the plant figured by Goeppert in his Foss. Flor. d. Perm. Form. Plate X, as *Sphenopteris integra.* The nervation agrees exactly with Brongniart's P. oreopteridia. We have not seen this plant at any lower horizon.

Pecopteris pennæformis, Brongt.; Var. *latifolia.* Pl. XVII, Figs. 4 and 5.

None of the typical forms of P. pennæformis have been seen in the upper measures of W. Va., but at Cassville we find with the Waynesburg coal a form which we consider a variety of this plant, and which we describe as Var. latifolia.

The pinnules of our variety are broader in proportion to their length than those of Brongniart's plant, and the nervation is somewhat different, as the lateral nerves make a greater angle with the mid-rib, in the pinnules. The facies is much like Heer's P. pennæformis, as figured in his Flor. Foss. Hel.

Pecopteris Miltoni, Artis, Pl. XXIII. Figs. 2 and 3.

On plate XXIII, Figs. 2 and 3, we give a form of this species which differs somewhat from all hitherto figured. It seems much more slender and narrow than the typical form. The mode of passing from entire pinnules, at the end of a compound pinna, through crenulate forms, into lobed ones, and finally into simple pinnules, is very gradual, and produces often very slender, elongate pinnæ and pinnules. Still this plant is so closely connected by transition forms with the typical one, that it cannot be separated even as a variety.

At Carmichael's, Penn., where this form occurs, we find immense numbers of this plant, which seems to exclude almost all other forms at this place. At West Union, in W. Va. also the plant is very common. The pinnæ often exhibit a spread of 2 and 3 feet, and fragments in great numbers of the stipes are found, many 5 or 6 inches in diameter.

At other localities we find the variety *polymorpha*, given by Brongniart as a distinct species (P. polymorpha.) This form usually occurs at different localities from those where the forms of Miltoni are abundant, and the facies of the plant is different, the pinnules being broader, longer, and less rounded at the extremity. Some splendid specimens of primary pinnæ, complete to the extremity, are found more than a foot long. Our forms of Miltoni agree precisely with those given by Artis, and have a somewhat different facies from that of Brongniart's plant.

Pecopteris dentata, Brongt. Pl. XXII, Figs. 1-5.

We find at Cassville, in the roof shales of the Waynesburg Coal, several forms of a plant which is so closely allied to this very polymorphous species that we do not think it proper to separate them further than as varieties. Figs. 1-4, Pl. XXII, show a well marked type, which is the most abundant form, and might be denominated *P. dentata, var. crenata*. The form figured in Fig. 4 exhibits some points of difference from that given in Fig. 1, in the pinnules being narrower and more constricted at the base, and more remote, and also in the tendency to become shorter towards the insertion of the ultimate pinnæ. The form given in Fig. 1 assumes more of the aspect of the typical plumosa form of P. dentata, especially in the lower pinnæ. Seen separately, the two plants might be taken as distinct species, or at least varieties, but we have so many intermediate forms at this place that no dividing line can be drawn between them.

The plant figured on plate XXII, in Fig. 2, differs a good deal from all the forms of the var. crenata above mentioned, and assumes the facies of the plumosa form of dentata. It differs however from Brongniart's plumosa in its more minutely dentate pinnules in the small size and delicacy of the pinnules, which, unlike the European plumosa, show no tendency to increase in size as we descend to lower pinnæ. It is still more widely separated from the form identified by some palæo-botanists as *P. plumosa*, from the Lower Coal Measures of America. This form might be distinguished by the varietal name *parva*.

We find a great number of well preserved forms of the true *dentata*, at Cassville, among which we find most of the forms figured by Brongniart, Geinitz and Heer; but we have as yet seen it at no other locality.

Pecopteris pteroides, Brongt.

This is one of the most widely distributed plants that we find in the Upper Carboniferous Strata, it being found at every locality where we have examined the flora of the Waynesburg Coal. Near Arnettsville, between Fairmont and Morgantown, in Monongalia Co., it is very abundant in the roof shales of this coal seam, and compound or primary pinnæ were seen 1½ feet long and a foot wide. Our plant has the facies and nervation of Germar's, given on plate XXXVI, in his Verst. d. Stein. Form. v. Wettin u. Löbj. At Carmichaels, Penn. it is very abundant.

Pecopteris Pluckeneti, Brongt. Pl. XXI, Figs. 4 and 5.

At West Union, in Doddridge Co. we find countless numbers of this plant, with every known and some new forms. Indeed the variableness of the plant is simply astonishing, and can be appreciated only when we have, as here, a great amount of well preserved material, which enables us to follow it through its many changes. Besides being thus abundant at this locality it is a widely diffused plant, for we find it at numerous other localities, in the Waynesburg Coal, as well as at all the higher horizons nearly to the top of the series. Figs. 4 and 5 give the most common forms of the plant as found at West Union, and it will be seen that though they do not differ essentially from some of the numerous types already figured by others, yet have a facies of their own. This plant must have been an arborescent species, from the great size which some of the specimens show. Some of the stipes are 5 or 6 inches in diameter, and fragments of fronds were seen 18 to 24 inches in length and width. The plant becomes much rarer as we ascend into the upper beds, and is much less abundant at other localities, in the Waynesburg Coal.

Pecopteris Pluckeneti. Brongt. var. *constricta.* Pl. XXI, Fig. 3.

Fig. 3, Pl. XXI, represents a species of Pecopteris which has many of the features of *P. Pluckeneti*, and yet differs from it in some important points. The general shape of the pinnules is different, since it is cut away or constricted at the base, which, being a constant feature, may determine the varietal name. The nerves also are different from those of Pluckeneti proper, since they branch more, and are more sharply defined and distinct from the parenchyma of the pinnule. It is possible that this may represent an entirely new species, and should it so prove, after the collection and examination of more and better material, it might then bear the name *Pecopteris constricta.* It occurs with *P. pluckeneti* in the roof-shales of the Waynesburg Coal, at West Union, Doddridge Co.

Pecopteris notata, Lesqx.

At the horizon of the Redstone Coal, near Wheeling, W. Va., we find a very beautiful little plant, which in the form of its pinnules, and in its nervation, cannot be distinguished from *P. notata,* as given by Prof. Lesquereux, in the "Geol. of Penn.," Vol. II, Part 2, Pl. XVIII, Fig. 4. It lacks the point-like dots which distinguish the plant of Lesquereux. This may however not be of specific value, since our plant resembles it so much in other respects.

Pecopteris Germari, (Weiss.) F. and W. Pl. XIX, Figs. 1–7.

Under the head of Cyatheites Pluckeneti, Weiss, in his excellent work on the "Fossile Flor. d. jün. Stein-kohlen-formation u. des Rothliegenden," describes a sub-species, *Cyatheites Germari,* and gives for it the following characters by which it is separated from Cyatheites Pluckeneti. "Pinnules pinnately parted, smoothe, contracted at the base;

lobes rotundate to sub-quadrate; nervation, as in C. Pluckeneti."

The form figured by Germar in his work, "Verst. d.
Stein," &c., Pl. XVII, Fig. 4, is referred by Weiss to the
same sub-species.

At West Union we find in the roof shales of the Waynesburg Coal abundant and beautifully preserved specimens
of a plant which agrees exactly with this description. Although it occurs with immense numbers of all possible forms
of P. Pluckeneti, yet it preserves always a distinct facies
which enables us at a glance to detect it, and no intermediate or transition forms are seen to indicate that it may pass
into P. Pluckeneti. It would seem then to be entitled to rank
as a distinct species. The characters may be given as follows:

Pecopteris Germari.

(Frond quadripinnate, elongate-elliptical in outline; secondary pinnæ, linear-lanceolate, inserted under an angle of
nearly 90°, stiff in aspect, with a broad flat rachis, which is
marked by a raised woody ridge on each border, and a
strongly striated, depressed central portion; tertiary pinnæ, linear to oblong, inserted at an angle of 45°, slightly
decurrent and in the lower portions of the frond cut into
from 5 to 7 pairs of rotundate, subquadrate or broadly
spatulate pinnules in the middle portions; these pinnules
pass into rounded lobes, which become less and less defined
until the pinnæ of this order pass in the terminal part of
the frond into pinnules; nerves diverging flabellately in the
pinnules or segments, being composed of lateral nerves
which fork once, and are nearly as strong as the middle
nerve.)

The parenchyma of the plant seems to have been thick
and dense, for it leaves a smooth shining film of carbonaceous matter. The nerves seem to have been imbedded in
the parenchyma, hence they are usually difficult to make
out. On macerated specimens they are seen to be rather
slender and sharply defined.

This plant, which is one of the most beautiful in the entire flora of the Coal Measures, has a considerable resemblance to Sphenopteris nummularia, Gutb., and approaches still more nearly to Pecopteris pinnatifida, (Gutbier) Gein. which, as is known, is a rare Permian plant.

It is found also at Cassville at the same horizon, and here the forms showing distinct pinnules on the ultimate pinnæ are more common. Pl. XIX, Fig. 6, represents a more finely divided form.

P. Germari, Variety crassinervis. Pl. XX, Fig. 5.

Fig. 5, plate XX, represents a form of Pecopteris so closely allied to P. Germari, that we have thought it best to consider it only as a variety of this species.   It is distinguished from the typical form by its very thick nerves, which are shown slightly enlarged in Fig. 5a.

The pinnules have also a crenulated border, and the whole plant differs somewhat from P. Germari.   Better and larger specimens may show it to be a distinct species.

It is found associated with P. Germari in the roof shales of the Waynesburg Coal, at West Union, W. Va.

P. Germari, Variety, cuspidata. Pl. XX, Fig. 4.

Figs. 4 and 4a, Plate XX, represents another form, allied to P. Germari, and found at West Union in the rich store of plants afforded by that locality.   This plant differs considerably from the forms on which the typical species is founded.

The lobes of the pinnules are tipped with sharp, rigid teeth, which, in part at least, are due to the prolongation of the nerves beyond the parenchyma of the pinnules. These are usually three or four in number in each lobe. The most divergent forms have but a slight resemblance to P. Germari, but there are so many intermediate forms that we cannot separate it as a distinct species.   It is found in the roof shales of the Waynesburg Coal at West Union, West Virginia.

Pecopteris sub-falcata. Sp. nov., Pl. XXI, Figs. 1-2.

(Frond bipinnate ; primary pinnæ large, triangular in out-

line, with a stout and rough rachis ; secondary pinnæ long,
narrow and pointed, alternate, departing from the main
rachis under an angle of 60° ; secondary rachis, terete,
straight, and rather strong ; pinnules, rounded at the apex,
slightly falcate, alternate, all inclined forward, or obliquely
inserted, and decurrent ; primary nerve of the pinnules
slender, but distinct ; secondary, or lateral nerves, diverging at an acute angle, forking once near the insertion, and
each branch, or only one, again forking before reaching the
margin.

This plant seems to stand about midway between *Pecopteris* and *Callipteridium*, for it has some of the features of
the latter genus in its nervation. The primary nerve however does not split up soon enough to form a true Callipteridium. The Pecopteris nearest allied to it is probably
P. pteroides, Brongt., from which it differs in its long,
pointed pinnæ, and also in the shape, insertion, and nervation of the pinnules. It resembles some of the Rhætic
Cladophlebids.

Habitat.—Roof shales of the Waynesburg Coal, Cassville,
West Virginia.

Pecopteris rarinervis. Sp. nov., Pl. XX, Figs. 6, 7 and 8.

(Frond bipinnate ; primary rachis slender and smooth ;
pinnæ alternate, linear-lanceolate, and going off at nearly
a right angle ; secondary rachis, slender and rigid ; pinnules alternate, short, ovate, rounded at the apex, united
for a short distance above the base in the lowest ones, and
becoming more united as we pass up towards the summit
of the frond, where the pinnæ pass into pinnules of linear
shape, with undulate margins : primary nerve of the pinnules distinct, and somewhat flexuous ; secondary nerves
few, passing off at an acute angle, and forking dichotomously.)

The nervation of this plant is similar to that of Pecopteris
Bredovi, Germar. The primary nerve is more distinct in
our plant, and the facies differs somewhat from that of
Germar's species.

Habitat.—Roof shales of the Waynesburg Coal, Cassville, West Virginia.

Pecopteris imbricata. Sp. nov., Pl. XXIII, Fig. 1.

(Frond tripinnate, and large, with a stout and rough primary rachis; pinnæ alternate, and going off at nearly a right angle, linear-lanceolate, and terminated by an obovate, or oblong-elliptical pinnule; pinnules apparently imbricated by the adjacent edges nearly to their summits. Very obtuse at the apex; middle nerve well defined, side nerves simple, and passing off at an angle of about 45°.

The basal pinnule on the lowest side is often inserted partly, and sometimes almost wholly, on the primary rachis.

This plant resembles very much *P. adiantoides*, L. & H. in some of its features, but differs from it in the imbrication of the pinnules, in the mode of departure of the lateral nerves from the median nerves of the pinnules, and in its more densely crowded appearance. The pinnæ themselves are often imbricated.

Habitat.—Roof shales of the Waynesburg Coal, Cassville, West Virginia.

Pecopteris asplenioides, Sp. nov., Pl. XXV, Fig. 1.

(Frond, tripinnate; primary rachis, strong and rough; pinnæ, close, and densely crowded, alternate, and going off at nearly a right angle; pinnule, alternate, ovate-oblong, and slightly contracted at the base, crowded closely together on the strong secondary rachis; primary nerves of the pinnules well marked, and extending to the apex; secondary or lateral nerves going off at an acute angle, forking once near the insertion, and each branch forking again near the margin of the pinnule; fertile pinnules on the same pinnæ intermingled with the sterile ones; fructification, arranged in two rows, composed of linear-elliptical sori which are inclined to the mid-nerve at an angle of about 60°, and extend from it to the margin of the pinnule.)

The sori appear to be placed on the lateral nerves. The resemblance of the fructification to that of Asplenium has given the name to the species. The intermingling of fertile and sterile pinnules is a rare feature.

Habitat.—Roof-shales of the Waynesburg Coal, Cassville, W. Va.

Pecopteris rotundifolia, Sp. nov.. Pl. XXIV. Fig. 6.

(Frond, tripinnate; primary pinnæ, lanceolate in outline. with a slender and somewhat flexuous rachis; secondary pinnæ, linear, alternate, and going off at nearly a right angle, with slender rachis; pinnules, short, rounded, united in the upper portion of the frond for some distance above their attachments, separate in the lower portions; mid-nerve, slightly flexuous, and not strongly marked; lateral nerves, passing off at an acute angle, forking once near the margin of the pinnule or lobe, and arching slightly upwards.)

Some forms of this plant have a slight resemblace to *P. concinna*, Lesqx. in the mode of nervation, but in our plant the lateral nerves fork, while in P. concinna they are mostly simple.

Habitat.—Roof-shales of the Waynesburg Coal, Cassville, W. Va.

Pecopteris platynervis. Sp. nov., Pl. XVIII, Figs. 1 6.

(Frond, tripinnate; primary rachis, strong, rough, and marked with pointed dot-like elevations; secondary pinnæ, alternate, linear-lanceolate, going off at nearly a right angle; secondary rachis, stout at the insertion. but tapering rather rapidly to the apex, where it is rather slender: pinnules, short, oblong, obtusely rounded at the apex, separate to the base in the lower and middle portions of the frond, becoming more and more united toward the summit, until they pass through pinnules with lobed and undulate borders finally into simple pinnules of the normal form; mid-nerve of the pinnules well marked and distinct to the apex; lateral nerves broad and flat, usually forking just at the point of insertion, thence diverging, without branching and almost without arching. to the margin. thus forming a

V shaped figure; in the lowest portion of the frond one branch again divides before reaching the margin.)

This plant varies a good deal in appearance according to the portion of the frond from which the specimen comes. It has always a peculiar rigid aspect. It is allied to *P. oreopteridia*, Brongt., but differs in the broad lateral nerves and their peculiar mode of diverging from the midnerve, and in the more gradual passage of pinnæ into pinnules toward the summit of the frond.

Figs. 6, 6a, 6b, Pl. XXVIII, show normal and magnified forms of a pinna and pinnules from the lower portion of the frond where the pinnules are larger and have the lateral nerves more complex than in the usual form. Pl. 18, Fig. 2b, shows the peculiar flat lateral nerves as seen under a lens when they are shown to consist of two consolidated bundles of nerves instead of one, as is usually the case in these lateral nerves. These two fibres, closely placed side by side, give the nerves their broad character. Fig. 1, Pl. 18, represents a segment of the middle portion of a primary pinna; Fig. 2 of the same plate a portion nearer the end, and Figs. 4 and 5 the extremity of the same pinna. Fig. 3 is probably a primary pinna near the summit of the frond.

The distribution of this plant is somewhat peculiar. At Cassville it is confined to the seam of shale which separates the highest layer of coal from the main mass, and has not certainly been seen above or below this 12 inch bed of shale. It occurs nowhere else, apparently, in the Upper Measures, but is found here in immense quantities.

Habitat.—Roof shales of the Waynesburg Coal, Cassville, West Virginia.

Pecopteris rotundiloba, Sp. nov., Pl. XVII, Fig. 2.

(Frond, tripinnate; primary rachis rather thick; secondary pinnæ, going off at an acute angle with a slender rachis; pinnules alternate linear with rounded lobes; primary nerve rather strong, and divided into nervules towards the end; nerves of the lobes mostly simple, and going off acutely from a well marked midnerve. In the terminal lobe, which is the largest, the midnerve divides dichotomously.)

Habitat.—Roof shales of the Waynesburg Coal, Cassville, West Virginia.

Pecopteris Schimperiana. Sp. nov. Pl. XXIV, Figs. 1-5.

(Frond, tripinnate; primary rachis stout, and rather rough; secondary pinnæ alternate, linear-lanceolate, taper-pointed, going off at nearly a right angle; pinnules, short-ovate, or triangular in outline, alternate, decurrent, and united near the base in the lower pinnæ, and more and more united as we pass towards the summit of the frond, or of the pinnæ; texture, thick and leathery; mid-nerve, strong and flexuous, extending to the apex; lateral nerves stout, and forking dichotomously in a straggling manner. The branches all being deflexed, so as to meet the margin of the pinnules almost under a right angle.)

This plant is one of the most distinctly characterized ones that we have met with in the Upper Carboniferous flora. It shows two forms, which present a somewhat different facies, viz: that given in Figs. 2, 1, 3 and 5, and the one depicted in Fig. 4. The first form has more acute pinnules, which, in small pinnæ near the summit of the frond, become quite pointed. The form given in Fig. 4, has obtuse, falcate pinnules. This if seen alone might be taken as a distinct species, or at least variety; but the peculiar nervation of the form first described is possessed by this also, and the presence of intermediate links forbid the separation of the two.

This species in several features closely resembles Brongniart's P. Sulziana, from the base of the Trias, as figured in the Hist. d. Veg. Foss., p. 225, Tab. CV, Fig. 4. It is possible that the change of facies seen in Fig. 4 is caused by the fact that this portion of the plant comes from the top of the frond. Fig. 2 shows a fragment from the lower part of the plant, where the triangular pinnules begin to show a tendency to become lobed, as if about to form new divisions. It is probable that still lower these may pass into pinnatifid pinnæ.

Habitat.—Roof shales of the Waynesburg Coal, West Union, West Virginia.

Pecopteris pachypteroides. Sp. nov., Pl. XXVI, Figs. 1–4.

(Frond, tripinnate; primary rachis, rather stout; secondary pinnæ, alternate, somewhat remote, having a narrowly winged, somewhat flexuous rachis; tertiary pinnæ (pinnules) numerous, alternate, very obliquely inserted on the rachis, and cut into 6–10 pairs of lobes; mid-rib broad and leather-like; lobes of the pinnules dense and coriaceous, with an indistinct mid-nerve, from which lateral nerves pass in a pinnate manner, but are obscurely shown, apparently simple.)

The texture of the pinnules is so dense, and the nerves are so deeply burried in the leaf substance, that the details of the nervation cannot be made out clearly. The lobes of the pinnules have a peculiar falcate or hooked form in the middle and upper part of the frond. In the lower portion they become crenate, as is shown in Fig. 1*a*.

Figs. 1*c* and 1*b* are enlarged pinnules from the middle and upper part of the frond, and show the form of the lobes there. Fig. 4 represents a detached terminal portion of a compound pinna.

Some of the forms of this plant have a strong resemblance to *P. dentata*, Brongt., but the plant is more finely cut and slender, while the decurrent pinnules and winged rachis are not found in P. dentata. It has a strong resemblance to *Pachypteris*, Brongt.

Habitat.—Roof shales of the Waynesburg Coal, Cassville, West Virginia.

Pecopteris angustipinna, Sp. nov., Pl. XXVII, Figs. 1–3.

(Frond, tripinnate; primary pinnæ, triangular in outline; primary rachis strong and arborescent; secondary pinnæ, elongate-linear, narrow, alternate and thickly set; secondary rachis stout and rigid; pinnules ovate-obtuse, slightly inclined forward or falcate, united at the base, the amount of union increasing toward the summit of the frond; parenchyma dense and leather-like; basal pinnule on the lower side, often partly inserted on the primary rachis; mid-nerve well defined; lateral nerves going off at an acute angle and forked.)

This plant resembles P. arborescens somewhat; but the union of the pinnules, the nervation, and the insertion of the pinnules distinguish it from that species. It was evidently a very fleshy plant, as the nerves are usually so deeply buried that they are seen with difficulty.

Habitat.—Roof shales of the Waynesburg Coal, West Union, West Virginia.

Pecopteris Heeriana. Sp. nov., Pl. XXV, Figs. 3–7.

(Frond, tripinnate; secondary pinnæ, alternate, somewhat flexuous, going off at nearly a right angle; pinnules, slightly falcate, remotely placed and decurrent on the rachis, so as to render it distinctly winged; lowest pinnule on the lowest side, often inserted partly on the principal rachis; pinnules on the lower portion of the plant, notched or lobed; mid-nerve, well defined and extending to the apex of the pinnule; lateral nerves, going off at an acute angle, forming 4 or 5 pairs, simple; fructification, composed of numerous shield-shaped sori covering the surface of the pinnule.)

The texture of the pinnules is thick and leather-like, usually obscuring the nerves. The plant presents a type unusual in the Carboniferous strata, but characteristic of the Rhaetic flora. It belongs to Schimper's section of Pecopteris acrostichides, and recalls forcibly the appearance of Pecopteris Williamsoni, Brongt., both in the form of the pinnules, and in the character of the fructification. Named in honor of Dr. Oswald Heer of Zurich.

Habitat.—Roof-shales of the Waynesburg Coal, Cassville, W. Va.

Pecopteris tenuinervis. Sp. nov., Pl. XXVIII, Figs. 1–4.

(Frond, tripinnate; primary rachis, strong, and somewhat rigid; secondary pinnæ, linear-lanceolate, alternate, closely placed, becoming gradually shorter towards the apex of the primary pinna, thus giving this a triangular outline; pinnules very short, narrow and alternate, the lowest one, on the lower side, always heteromorphous, it having a crenulate margin, and being larger than the rest; pinnules toward the lower part of the plant, all longer and larger than the normal ones, and possessing rounded lobes; pinnules of the middle portion (normal pinnules), small, oblong, obtusely rounded at the end; pinnules near the summit of the same shape, but very minute ; mid-nerve, well-defined, but slender ; lateral nerves, all very delicate, those in the lower lobed pinnules twice forked, those in the central portion of the plant once forked a short distance above their insertion ; fructification, consisting of two rows of rounded or slightly elliptical sori, raised like mamillæ. placed on each side of the mid-nerve, and covering the greater portion of the surface of the pinnule.)

The texture of the plant seems to have been pretty dense, and the compression of the slender nerves in this thick substance causes them usually to have a peculiar entangled appearance. The sori are so closely placed, that they often appear to be imbricated. Fig. 1 represents the pinnules from the lower part of the plant, where they appear to tend to pass into pinnæ. The general facies of the plant resembles the more delicately cut forms of P. arborescens, but the points of difference are well marked and constant.

Habitat.—Roof-shales of the Waynesburg Coal, Cassville, W. Va.

Pecopteris Merianiopteroides. Sp. nov., Pl. XXIX. Figs. 1-2.

(Frond, tripinnate; primary pinnæ, triangular in outline; secondary pinnæ, linear-lanceolate, going off at almost a right angle ; pinnules, obtusely ovate, united at the base, and inclined slightly forward ; mid-nerve well defined, lateral nerves numerous, once forking, and departing under an acute angle, those from the lower side of the pinnule passing off from the attachment of the pinnule to the rachis.)

The general facies of the plant, together with its nervation, very much resemble the form described by Heer in his "Trias u. Jura. Pflanzen," on which he founded the new genus of Merianiopteris, hence the name we have given it.

Habitat.—Roof shales of the Waynesburg Coal, Cassville, West Virginia, and Carmichiael's, Pennsylvania.

Pecopteris ovoides. Sp. nov., Pl. XXIX, Fig. 3.

(Frond, tripinnate; primary pinnæ, tapering rapidly to the summit; secondary pinnæ, alternate, placed thickly, and going off from the primary rachis at an acute angle: pinnules, ovate, united at the base; mid-nerve, strong; lateral nerves, making very acute angle with the mid-nerve, forming 6 or 7 pairs only, simple.)

Habitat.—Chocolate shales, 400 feet above the Waynesburg Coal, Bellton, Marshall county, West Virginia.

Pecopteris lanceolata. Sp. nov. Pl. XXIX, Figures 7, 8 and 9.

(Frond, tripinnate; secondary pinnæ, alternate, somewhat crowded; pinnules, lanceolate, united for some distance, and curving slightly forwards; mid-nerve distinct; lateral nerves few in number, going off almost at a right angle, simple.)

This beautiful little plant has some resemblance to Pecopteris Unita, Brongt., but the pinnules are more delicate than those in that species, and have a characteristic forward inclination not seen in P. unita.

Habitat.—A shale at Bellton, 400 feet above the Waynesburg Coal; and at Moundsville, West Virginia, at the horizon of the Waynesburg Coal.

Pecopteris latifolia. Sp. nov., Pl. XXIX, Figs. 5-6.

(Frond, tripinnate; principal rachis strong; secondary pinnæ very closely set, alternate; secondary rachis very stout; pinnules united at the base, broad, bluntly ovate: mid-nerve well marked; lateral nerves going off at an acute angle, and once forking near the insertion.)

This plant has a large vertical range, as we find it at Cassville, in the roof shales of the Waynesburg Coal, and also at Bellton, West Virginia, 400 feet above the Cassville horizon.

Pecopteris inclinata. Sp. nov., Pl. XXIX, Fig. 4.

(Frond, bi-trippinnate; pinnæ small and delicate; pinnules bluntly lanceolate, separate, alternate, and inserted on the rachis at a very acute angle ; mid-nerve well defined, and extending to the apex ; lateral nerves few, simple, and going off at an angle of 45°.)

Habitat.—Roof shales of the Waynesburg Coal, Cassville, West Virginia.

Pecopteris, Species? Pl. XXVII, Fig. 5.

Fig. 5. Pl. XXVII, represents a form of Pecopteris that we find in fragmentary specimens at Cassville, in the roof shales of the Waynesburg Coal, associated with *Goniopteris emarginata*, Schimp., which in the texture and tapering nature of pinnæ somewhat resembles what we might suppose the lower pinnæ of this Goniopteris to be; but the forking lateral nerves of the pinnules, their slight union, and falcate form show that it is a different plant. It is very probably a new species, but as we have not seen any larger specimens than the one figured, we cannot fix with sufficient certainty its specific character, and hence forbear to give it a name.

Pecopteris Species? Pl. XXVII, Fig. 4.

In Fig. 4, Pl. XXVII, we give a small fragment of a plant which very much resembles *P. sub-falcata*, F. & W., but the insertion of the pinnules is quite different from that in the latter plant, and the mid-nerve, which in *P. subfalcata* is rather slender, is here very thick, especially towards its base. It may represent a new species, but as yet we have not sufficient material to fix its specific character.

Habitat.—Roof shales of the Waynesburg Coal, Cassville, West Virginia.

Pecopteris goniopteroides. Sp. nov., Pl. XXV, Fig. 2.

(Frond, tri-pinnate; principal rachis very stout; secondary pinnæ alternate, narrow, almost linear, going off at nearly a right angle ; secondary rachis quite slender ; pinnules united to near the middle, ovoid, inclined forward ;.

median nerve slender, but distinctly marked; lateral nerves ascending under a very acute angle, producing a flabellate nervation, forking once near the middle of the pinnule, the two lowest adjoining pairs from each pinnule arching up abruptly towards the sinus, so as to leave triangular spaces destitute of nerves, as in Goniopteris, or Cynoglossa, to both of which genera it has a strong resemblance.)

Habitat.—Roof shales of the Waynesburg Coal, Cassville, West Virginia.

Pecopteris Sp?   Pl. XXIV, Fig. 7.

This beautiful little fragment is well marked, and distinct from all species known to us, so far as the portion of the frond shown in the specimen can determine this.   The pinnæ are slender, obliquely placed, and cut into rounded ovate lobes, which are directed forwards.   The free lower margin of the lobes is much longer than the upper; the rachis of the pinnæ is very slender; the lobes have a slender mid-rib, furnished with simple lateral nerves, which go off so as to be directed toward the end of the pinnule.

The plant resembles a Goniopteris.   Though so well marked we have thought it best not to fix the species on so small a fragment.

Habitat.—Roof shales of the Waynesburg Coal, Cassville, West Virginia.

GONIOPTERIS. Presl, emend. Al. Braun.

The plants of the genus Goniopteris, as limited by Schimper, are among the most characteristic ones of the Upper Carboniferous. They are in some localities very abundant, and form one of the features by which the flora of the higher strata is distinguished from that of lower horizons. No species of the genus is found in West Virginia below the horizon of the Pittsburg Coal. It occurs in all horizons from the Waynesburg Coal up to 800 feet above it, and near the top of the Upper Barren Group.

6 PP.

Goniopteris emarginata (Goepp) Schimp.

This species was found by Bunbury in the Frostburg Coal Basin. We have in our remarks on Pecopteris elliptica stated that it is probable that the horizon of the Waynesburg Coal is exposed in that Basin in Maryland. It is found in W. Va. throughout the entire thickness of the Upper Barrens, at the following localities: roof-shales of the Waynesburg Coal, Cassville, W. Va. and red shales at Bellton, Marshall Co. 800 feet above the Waynesburg Coal. The plant is not quite so large as Goeppert's, but in other respects it is identical. The pinnules of our plant, and especially the forms found at the Bellton locality, are shorter than those of the typical species.

Goniopteris elegans, (Germ.) Schimp.

A few fragments of this plant have been seen in the roof-shales of the Waynesburg Coal at Cassville, and their identity with Germar's species is unquestionable, since the fragments in question are distinct, and almost fac-similes of the typical plant.

Goniopteris longifolia, (Brongt.) Schimp.

A few fragments of this beautiful little plant were recognized in the roof-shales of the Waynesburg Coal at Cassville, W. V. A detached pinna 6 inches long was seen, in which the end was not preserved. The parenchyma was evidently thick and leather-like, and the specimens have a smooth shining appearance.

Goniopteris argula (Brongt.) Schimp.

This species is quite abundant both at Cassville and at West Union, in the roof shales of the Waynesburg Coal. It is slightly changed from the type given by Brongt., having somewhat longer pinnules or segments, which taper rather more towards their apex. The nerves are also rather stronger than those of the typical plant. Our plant rather resembles Geinitz's figures for this species, given in his Stein. v. Sachs. than those of Brongt.

Goniopteris elliptica, Sp. nov., Pl. XXX, Fig. 1.

(Frond, bipinnate; pinnæ closely placed; rachis slender and somewhat flexuous; pinnules alternate, narrow, elliptical, and somewhat acute, united too near the middle; midnerve well marked, slender and extending to the apex; lateral nerves simple, about 6 on a side, the lowest pairs of adjacent nerves usually meeting at an acute angle, and all going off at an angle of somewhat less than 45°.)

Habitat.—Roof-shales of the Waynesburg Coal, Cassville, West Virginia.

Goniopteris Species? Pl. XVII, Fig. 6.

This fragment of a pinna was found in the Roof-shales of the Waynesburg Coal at Cassville, W. Va. and resembles the plant figured by Prof. Lesquereux on Plate XIII, Fig. 12 Illinois Report, and referred by him to *Pecopteris (Goniopteris) arguta*. Our plant is fruiting, as is that of Prof. Lesquereux. This plant does not agree with the typical form of *G. arguta*, which we find at the same locality, and it is probable that it should be referred to a new species.

Goniopteris oblonga, Sp. nov., Pl. XXX, Figs. 3–5.

(Frond, bipinnate; primary rachis rough and stout; pinnæ toward the base of the frond alternate, closely placed, going off at angle above 45°, arching downward slightly, with a rigid aspect, linear-lanceolate; pinnules, alternate, crowded, rounded at the apex, united to near the middle; in ascending, more and more united; terminal pinnules, united to the summit; lowest pinnule on the lower side, more or less deflexed; pinnæ toward the summit of the frond, becoming shorter in ascending, with lobes less and less defined, until they pass into pinnules with undulate margins; mid-nerve strong and extending to the apex of the lobe; lateral nerves simple, going off at an acute angle, the lowest pair anastomosing with the corresponding pair of the adjacent lobes at the sinus.)

Habitat.—Roof-shales of the Waynesburg Coal, West Union, W. Va.

Goniopteris Newberriana, Sp. nov., Pl. XXX, Fig. 2.

(Frond, tripinnatifid; primary pinnæ, triangular in outline, tapering rapidly toward the apex; rachis, rather stout and rigid; secondary pinnæ (pinnules), alternate, linear-lanceolate, narrow, closely placed and going off at almost a right angle with the primary rachis, cut into numerous ovate-acute segments or lobes, which are minutely dentate and become narrower, more acute and more united toward the summit of the pinnæ; mid-nerve, well defined; lateral nerves, passing off into each segment or lobe, from which branches proceed in a pinnate manner, one into each tooth, the lower pair of branches proceeding to meet the corresponding branches of the adjacent segments at the sinus, but not uniting with them.)

This beautiful and finely cut plant has a thick coriaceous leaf substance which leaves a shining film on the stone. It differs slightly from the typical Goniopterids in the lowest pair of nerves failing to anastomose with their neighbors, but its features in all other points corresponds so fully with those of the genus that we do not feel justified in separating it from *Goniopteris*. It is much like *Pecopteris arguta* of Brongt., Schimper's *Goniopteris arguta*, but much more finely cut, and is also smaller. It is named in honor of Dr. J. S. Newberry, the distinguished palæobotanist of Columbia College.

Habitat.—Roof-shales of the Waynesburg Coal, West Union, W. Va.

CYMOGLOSSA, Schimper.

The genus Cymoglossa was founded by Schimper on the *Pecopteris Goepperti* of Morris, a plant from the Permian of Russia. According to Schimper it includes plants with the facies of Goniopteris, but having the tertiary or ultimate nerves of the lobes or pinnules in large part forked. He gives the following as the generic character:

"Frond pinnate; pinnæ, broadly oblong, or elongate-lingulate, undulate (whence the name; *glossa*, tongue, *kuma*, wave) margin, with short round lobes. Nerves of the united pinnules, leaving the rachis of the pinnæ under an acute angle, arcuate-diverging, alternate; secondary nerves of the pinnules (tertiary of the pinna), arising at a very acute angle, numerous, all verging towards the margin of the pinna, the two lowest anastomosing at the sinus of the lobules with their neighbors, simple, forming with the rachis a long triangle destitute of nerves, the others reaching the margin of the lobule, simple, and forked."

As this generic character is based on a single species, and since we have several plants which have the essential features of the genus, but differ in the nervation from Goniopteris too much to be included in the latter genus, we think it proper to amend the generic character, as given by Schimper, so as to include the plants found by us. As amended we would have the following:

Frond, pinnate or bipinnate; pinnæ, linear-elongate, or elongate-oblong; undulate, or pinnatifid; mid-nerve of the united pinnules, leaving the rachis under an acute angle, alternate; lateral nerves, rising at a very acute angle, all verging upwards towards the margin of the pinnules, the two lowest uniting with or meeting the corresponding ones of the preceding and following pinnules at the sinus of the lobes, simple or forked, and forming with the principal rachis a triangular area destitute of nerves, the others reaching the margin of the lobes, simple or forked.

Cymoglossa obtusifolia. Sp. nov., Pl. XXXI. Figs. 5 6.

(Frond, bipinnate; pinnæ, long, narrow, and tapering gradually to the summit, sessile, with a cordate appearance at the base, produced by the projecting downwards of the lowest pair of lobes or united pinnules; rachis, rather strong and pilose; pinnules, ovate or elliptical, obtuse, united to near the apex, rather fleshy, the lower pair heteromorphous, larger than the normal ones, and slightly deflexed; primary nerves, distinctly marked, slender; lateral nerves, very distinct, but slender, leaving the median nerve under an acute angle, arching upwards towards the margin of the pinnules, normally simple, but frequently forking; the lowest pair, simple, anastomosing with the corresponding ones of the adjoining pinnules, forming triangular spaces devoid of nerves; lateral nerves of the heteromorphous lower pinnules, more complex than on the lower side of the pinnules, forking occasionally twice, and all on the upper side once forking.)

The facies of this plant is much like that of *Goniopteris emarginata* (Goepp.) Schimp. but the forking nerves remove it from that genus, as limited by Schimper.

Habitat.—Roof shales of the Waynesburg Coal, Cassville, West Virginia.

Cymoglossa breviloba.   Sp. nov., Pl. XXXI, Fig. 3.

(Frond, bipinnate ; pinnae short, oblong, sessile, and slightly contracted at the base, alternate, inserted at a right angle with the primary rachis; margin slightly lobed, or only undulate; nerves passing off very acutely in groups into the segments or pinnules, which by their union compose the pinnules or pinnae, all reaching the margin, forking, and simple, the two lowest anastomosing with the corresponding ones of the adjoining segments at the sinus of the lobes, and forming long, curved, triangular areas without nerves.)

This beautiful fern corresponds closely with the typical plant of Schimper, *Cymoglossa Goeppertiana*, but is a smaller plant, and the one of the anastomosing nerves nearer the end of the pinnule is generally forked and unites with its neighbors by one branch. The texture is dense and leather-like, and the nerves, though rather slender, are very distinct and sharply outlined.

Habitat.—Roof shales of the Waynesburg Coal, Cassville, West Virginia.

Cymoglossa formosa.   Sp. nov., Pl. XXXI, Figs. 1-2.

(Frond, bipinnatifid ; pinnae long, linear, and tapering slowly to the extremity; rachis, rigid, rather slender, and marked by a raised, cord-like line along each margin; united pinnules or segments, oblong-lanceolate, terminating acutely, and dentately lobed ; mid-nerve of the segments going off at an acute angle, somewhat arcuately diverging, strong and rigid, extending to the apex; lateral nerves thick and rigid, leaving at an acute angle, verging upwards and passing into each tooth, forking near the extremity, the two adjoining lowest ones of adjacent segments meeting abruptly, and interlacing, forming the usual triangular space without nerves.)

This plant has a close resemblance with *Goniopteris arguta* (Brongt.) Schimp., especially the plant figured for this species by Geinetz, in his Steinkohl. von Sachs. but its strong forking nerves and thick parenchyma distinguish our plant. In Fig. 2, Pl. XXXI, we depict a pinna as found near a fragment of a stem, which is most probably a portion of the primary rachis to which were attached the isolated pinnae, which are the only forms found. This fragment is stout, rigid, and smooth, agreeing well with what we would expect to be the rachis of a primary pinna.

Habitat. — Roof shales of the Waynesburg Coal, Cassville, West Virginia.

Cymoglossa lobata, Sp. nov., Pl. XXXI, Fig. 4.

(Frond, simply pinnate; rather slender and delicate; pinnules oblong, crenately lobed, or undulate; primary nerve strong, and distinctly marked; lateral nerves passing off at an acute angle, and branching dichotomously, so as to form a flabellate group in segment of the pinnule, the lowest branch on adjacent sides of two groups meeting at the sinus near the margin of the pinnules and forming triangular areas without nerves.)

Habitat. — Roof shales of the Waynesburg Coal, Cassville, West Virginia.

ALETHOPTERIS, Sternb.

This genus is remarkable for the rarity of its occurrence in the Upper Carboniferous. We have seen but two species above the Pittsburg Coal, and this at only one locality. In the flora of the Lower Productive Measures, as well as in that of the Conglomerate group, this is one of the most abundant forms, forming almost the entire flora, as in the Sharon Coal of the Conglomerate of Pennsylvania. In species in the upper strata we have to note a total change of facies from the coarse large forms with strong and sparingly forked nerves found at lower horizons, to the type which approaches close to Callipteridium in nervation, while it is more slender, and shows a tendency to heteromorphism.

Alethopteris Virginiana, Sp. nov.. Pl. XXXII, Figs. 1–5. Pl. XXXIII. Figs. 1–4.

(Frond, tripinnate: primary rachis strong and rough; primary pinnæ triangular in outline, and tapering rapidly to the summit; secondary pinnæ. opposite or alternate, going off at nearly a right angle, long and tapering slowly, with a large and rather rigid rachis; pinnules, alternate, separate below but united above, and becoming more so as we approach the summit of the primary pinnae where the pinnules have all united, and the ultimate pinnæ are reduced to long, undulate or lobed pinnules, which finally pass into simple pinnules of the normal kind; the pinnules also coalesce towards the ends of the ultimate pinnæ, and are often swollen at the base, as if by two sori, placed one on each side of the mid-nerve at the base, as shown in Fig. 1. Pl. XXXIII; mid-nerve well marked, and extending to the apex; lateral nerves numerous, closely placed. going off nearly at right angles with the mid-nerve. Very fine, forking once normally, or with one of the branches, (occasionally both) again forking, simple nerves occasionally interspersed. all proceeding nearly parallel to each other to the margin: lowest pinnule on each side of the base of the pinna. of the ultimate order, heteromorphous by having the lower side of the pinnule lobed while the upper side is entire.)

Fig. 2, Pl. XXXIII, shows a form from the lower part of the plant, where the pinnules have a tendency to become lobed; and Fig. 2 shows this lobing in a more decided manner, thus causing the plant to tend to a quadripinnatifid character.

The pinnæ of the ultimate order in this plant were very long, for we have seen them incomplete and yet more than a foot in length. They must also have been very deciduous, for we find almost always only detached pinnæ. They lie by thousands in the shale, forming often all the plants found in a particular layer.

The distribution of this plant is very peculiar at Cassville where it occurs. The Waynesburg Coal is divided into three benches, by two partings of shale, one near the middle and the other near the top of the bed, and above this last or second parting there is usually about 12 inches of coal. In the shale under this top or "roof coal," is the habitat of our plant. The shale itself is usually about 12 inches thick, of fine grain and well adapted to the preservation of plants.

The Alethopteris occupies this shale, and excludes almost entirely all other plants. Above the "roof-coal," in the roof-shales, where we find nearly all our other plants from this locality, we never find the Alethopteris, either here or elsewhere. It seems extinguished in the subsidence causing the deposit of this shale. The plant is very polymorphous, so much so indeed, that but for the abundant material afforded, which enables us to obtain a number of intermediate forms, we would have been tempted to form several species out of this one. Fig. 1, Pl. XXXIII, gives an enlarged form of the pinnules with swellings at the base, which we take for fructifications. Prof. Lesquereux, in the Illinois Report, Vol. IV, Pl. 10, Fig. 6, gives a similar form of fructification, as shown in his *Alethopteris inflata.*

Alethopteris gigas, Gein. Plate XXXIII. Figs. 5 and 6.

We give on Pl. XXXIII, in Figs. 5 and 6, a representation of a plant which in its general appearance cannot be distinguished from *A. gigas,* Gein. It has the same shaped pinnules, the same large and swollen looking mid-nerve of the pinnules, and the same general facies. The plant is found only in sandy shale, which does not preserve its lateral nerves, hence we cannot identify it positively with Geinitz's plant.

We find near Bellaire, Ohio, 20 feet below the Pittsburg Coal, a plant which resembles the one found in the Upper Barren Measures, but it is larger and stouter in every respect. In this the lateral nerves are preserved, and are coarse and single, or once forked; hence this is not A. gigas of the Permian. The resemblance of this plant to the one now in question throws some doubt on the identity of the Upper Barrens' plant with the Permian form. But for the possibility that the Bellaire species has ascended into the Upper Strata, we should have no hesitation in identifying the plant at the higher horizon with *A. gigas.*

Habitat.—Sandy shale, at Bellton, Marshall Co. 500 feet above the Waynesburg Coal.

Taeniopteris, Brongt.

The finding of Taeniopterids with a well marked Permian facies among the plants of the horizon of the Waynesburg Coal, is a most significant indication of the important changes which the flora of the Carboniferous upper strata have undergone when compared with that of the horizons below the Pittsburg Coal. No plants of such a type have been found at any lower horizon.

A still more interesting feature is the discovery of fruiting forms of this genus, which show the character of the fructification, hitherto unknown, in the most unmistakable manner.

Schimper has separated the genus *Oleandridium* from Taeniopteris, taking apparently as his type species *Taeniopteris Vittata,* Brongt. He gives no reason for separating T. Vittata from the rest, or for founding a distinct genus "Oleandridium." Had he defined this genus better we would perhaps find ourselves compelled to place our fruiting plant in it, as this form is much like Oleandra in form and nervation, and besides, possesses a fructification not unlike Oleandra, in position at least, and arrangement. We place all our forms provisionally in the genus Taeniopteris.

Taeniopteris Lescuriana, Sp. nov., Pl. XXXIV, Fig. 9.

(Frond, simple, broad, elongate ; mid-rib, rather strong and rough ; lateral nerves, rather remote, somewhat numerous, going off from the mid-rib at an acute angle, forking once near the insertion, each branch usually forking again a short distance from the mid-rib, arching strongly outward so as to pass to the margin at right angles to it ; sometimes the branches fork near the margin, but rarely.)

As will be seen from the figure, the specimen given is only a fragment in which the margin of the part seen is not preserved. The part of the lamina preserved is 4 c. m.'s wide, and the entire plant must have been 10 cms. wide, as not one half is shown in the figure.

The nearest relative of our plant seems to be *Taeniopteris multinervis* of Weiss, Flor. d. jünst Steink. u. d. Roth. Tab. 81, Fig. 13.

Our plant however is larger, has a more slender mid-rib and few lateral nerves, with a different mode of forking, though the departure from the mid-ribs is similar.

Our plant resembles in size and form the Macrotaeniopterids of the Rhaetic and Oolite, and may be the ancestor of those found in the Richmond coal field.

It has a remarkably strong resemblance to Macrotaeniopteris (Schenk) gigantea of the Rhaetic, as figured by Schenk in his " Foss. Flor. d. Grensch." Pl. XXVIII, Fig. 12, both in the nervation as shown in the lower part of Schenk's figure in the size of the mid-rib and in the probable dimensions ; for in our plant the width could not have been much less than that of *M. gigantea*.

Our plant is named in honor of Prof. Leo Lesquereux, who has done so much to advance the science of palaeobotany.

Habitat.—Roof shales of the Waynesburg Coal, Cassville, West Virginia.

Taeniopteris Newberriana, Sp. nov., Pl. XXXIV, Figs. 1-8.

(Frond, simple, elongate, narrowly elliptical, tapering slowly to the apex and base ; mid-rib, of medium size, tapering gradually from the base to the apex of the frond ;

sterile fronds, about 2¾ cms. wide, and 20 cms. long; leaf-substance, rather thick and coriaceous, having a smooth, shining, carbonaceous film; lateral nerves, very fine, closely placed, and immersed in the parenchyma of the frond, leaving the mid-rib at a right angle, or with a very slight arch immediately at the insertion, mostly simple, but frequently branching once at irregular distances from the mid-rib, more rarely one or both of the branches again branching, all in a peculiar dichotomous manner, so that the nerves and branches continue parallel to each other to the margin; fertile frond, usually much smaller than the sterile one, and narrower, entire near the base, cut into segments which extend about half way to the mid-rib in the middle, and upper part of the frond; segments separated by very acute angled sinuses, round to truncate at the extremity, void of nerves, and containing beneath the sinus oval sori, which are apparently attached by their broad base to a receptacle near the mid-rib; receptacle, elliptical, flattened on one side, and leaving on each side of the mid-rib a row of distinct impressions of the same shape.)

The fertile frond contains two rows of broadly ovate sori, which stand one on each side of the mid-rib, and are so placed that the axis of each sorus stands perpendicular to the mid-rib, and just under the sinus separating each pair of segments. The sori extend just up to the bottom of the sinus. The basal portion of the fertile frond is free from segmentation and fructification, and possesses nerves like the sterile frond. The segmented portion shows no nerves. The segmentation of a frond often begins before the appearance of the sori, as shown in the plant given in Fig. 7. Fig. 2 gives the normal form of the fruiting frond. Fig. 1 is the middle portion of a fertile frond. Fig. 3 shows the plant with the impressions left by the insertions of the base of the sori. 3a gives an enlarged representation of the impressions, and Fig. 1a of the sori, with their bases at the upper part of the figure. Neither the sterile nor the fertile fronds have been seen entire. Figs. 4, 5, and 6 give the base, middle portion, and end of the sterile frond.

Macrotaeniopteris Rogersi. Schimp. of the Richmond

coal field, contains, on specimens in our possession, elliptical depressions strikingly like the depressions seen on this plant, and shown in Plate XXXIV. Fig. 3. In the specimen from the Richmond coal the depressions are larger, and are placed in one row *on* the mid-rib. Prof. Wm. B. Rogers, however, in his description of this plant, says they often occur in two rows, one on each side of the mid-rib. This form of fructification in Taeniopteris Newberriana, and the facies of T. Lescuriana, show that these Taeniopterids are probably the ancestors of the Macrotaeniopterids of the Mesozoic.

Our plant has a very considerable resemblance to *T. coriacea*, Goeppert, but is larger. It also resembles *T. vittata*, Brongt., in nervation and general form, but the mid-rib is flatter and more delicate. In general form, nervation, and in the position and arrangement of the sori, this plant is strikingly like *Oleandra nereiformis*, Presl., from the Island of Luzon, and this resemblance might call for the placing of it in the genus *Oleandridium*, Sch. if this were more distinctly defined.

The segmentation of the fertile frond has a curious resemblance to the pinnules of Pterophyllum, a plant which makes its appearance with well marked features in the Trias.

Taeniopteris Newberriana, Var. angusta. Pl. XXXIV, Fig. 8.

We find with the normal broad form a narrower and smaller frond, which is also seen in fructification. This, in all points except size, is similar to the larger plant. It may perhaps be placed as a variety under *T. Newberriana*.

RHACOPHYLLUM, Schimp.

Rhacophyllum filiciforme, Var. majus. Pl. XXXV. Fig. 1.

The plant figured in Fig. 11, Pl. XXXV, resembles the one figured by Schimper, in Pal. Veg. Tab. XLVIII, Figs. 3–6, so much in its general aspect, that we consider it as only a variety of R. filiciforme. It is much larger than Schimper's, and seems to possess more woody material in its ribs. Our plant is found associated with Pecopteris tenuinervis, though it has never been seen attached to any plant. The frond and segments show no distinct nerves. Along the axis of the plant we find a sort of woody rib, which sends obscure ribs into the lobes, which dissolve in striations. The texture is fleshy. The Eremopterid facies and obscure nervation cause the plant to resemble the fine *Gleichenites Neesii* of Goeppert, from the Permian of Europe.

Schimper figures his plant as attached to the stipe of P. dentata.

Habitat.—Roof shales of the Waynesburg Coal, Cassville, West Virginia.

Rhacophyllum laciniatum. Sp. nov., Pl. XXXV, Fig. 2.

(Frond, simple, smooth, tapering rapidly to the base or point of attachment, and presenting a cuneate outline ; laciniae, numerous, not deeply incised into the frond, and mostly simple, but in some cases again cut into segments ; nervation, not very distinct, nerves diverging flabellately from the base, forking frequently, branches passing into the laciniae.)

This plant is most nearly allied with R. filiciforme (Gutb.) Schimp. but is less coriacious, and the nerves are more distinct. It is found attached to *Pecopteris dentata*.

Habitat.—Roof shales of the Waynesburg Coal, Cassville, West Virginia.

Rhacophyllum lactuca, (Sternb.) Schimp.

In the roof shales of the Waynesburg Coal, at Cassville, West Union, W. Va. and at Carmichael's, Penn. we find large specimens of this plant. It seems to have been quite fleshy.

Rhacophyllum speciocissimum, Schimp. (Schizopteris lactuca, (Presl.) Roehl.)

We find several specimens of this splendid plant at Carmichael's, Penn. in the roof shales of the Waynesburg

Coal, some of them 8 inches long, and 6-8 inches wide, indicating a form fully as large as the fine plant figured by Von Röhl in his Foss. Flor. von Westp., Tab. XVIII.

CAULOPTERIS, Lind et Hut.

Caulopteris elliptica. Sp. nov., Pl. XXXV, Figs. 4 and 5.

(Scars large, arranged in quincunx order, mostly elliptical in outline, but some approaching an oval form; outer surface of the bark ornamented by irregular pits, and punctate elevations, perhaps from the insertion of aerial rootlets.)

Fig. 3 represents a single isolated scar which was found unconnected with others. It possesses a somewhat different shape from those shown in Fig. 4, being oval, and it is also somewhat larger; it may belong to a different species, but as it has so much in common with those of Fig. 4, we do not separate them.

Habitat.—Roof-shales of the Waynesburg Coal, Cassville, W. Va.

Caulopteris gigantea. Sp. nov., Pl. XXXVI, Fig. 1.

(Caudex, rough, very large, and furrowed; cicatrice, very large, broadly elliptical, not confluent at the extremities; vascular bundles, producing longitudinal furrows, and causing a roughened or broken appearance at the extremities of the scars, or sometimes near their centers; outer surface of the bark, ornamented with rounded pits and elevations Fig. 5, Pl. XXXV.

This species is more closely allied with *C. macrodiscus*, Corda, than with any other hitherto described species; but it differs from *C. macrodiscus* in the different shape and larger size of the scars. Both have smooth bordering spaces running around the scars like a frame. In our plant these are seen to have on the outer surface the markings given in Fig. 5. The smooth borders are apparently caused by impressions of the inner side of the bark surrounding the scars. It is probable that this may be the caudex of the fern which in its fronds gives us the forms of Alethopteris Virginiana, as the two are always associated in the roof shales.

Specimens of this Caulopteris have been seen more than 1½ feet broad.

C. gigantea, Stipes. Pl. XXXVII, Fig. 5.

This figure represents certain forms which we find, by the hundred, in the same shale with Alethopteris Virginiana and Caulopteris gigantea. They are of varying lengths and have sometimes the thickness of 2 or more inches. They seem to be impressions of the bark of fern stipes, and may belong to C. gigantea.

SIGILLARIA, Brongt.

Sigillaria approximata, Sp. nov., Pl. XXXVII, Fig. 3.

(Leaf scars, very ornamental, hexagonal, horizontal diameter nearly twice as long as the vertical, and terminating in acute angles at the extremities of the longer diameter, closely approximate ; decorticated stem, marked by longitudinal furrows, one between each row of leaf scars ; vascular scars, thin, the middle one slightly concave above and convex below, larger than the lateral ones, and transversely elongated, the two lateral scars are placed slightly above the middle scar, one at each end, and are punctiform and much smaller than the middle scar.)

This plant belongs to the Sigillaria of the type of S. Menardi, a form characteristic of the upper portions of the Carboniferous system everywhere. It is the only Sigillaria except S. Menardi that we have seen in the upper beds above the Pittsburg Coal in West Virginia. It is very rare, only two specimens having been seen.

Habitat.—Roof shales of the Waynesburg Coal near Arnettsville, W. Va.

Sigillaria Brardii, Brongt.

This species has not been seen in W. Va., but near Washington, Penn. it is very abundant in the roof of the Washington coal.

CORDAITES, Ung.

Cordaites crassinervis, Sp. nov., Pl. XXXVII, Fig. 10.

Fig. 10, Pl. XXXVII, seems to represent a species of Cordaites quite different from any hitherto described. The fragment has a very tapering form, and is somewhat thick and coriaceous. The nerves are very large and coarse, and are seen to branch again and again in leaving the point of attachment or base of the leaf. This plant may not belong to *Cordaites*, it may possibly represent a *psygmophyllum*. Not enough is shown to determine this point.

Habitat.—Roof-shales of the Waynesburg Coal, Cassville, W. Va.

Genus? Pl. XXXVII, Fig. 4.

We have given in Fig. 4, Pl. XXXVII, a very curious looking plant, of which we have not found more than one specimen, which is not sufficient to fix its generic position. It is flabellate in outline, and possesses rigid looking ribs which diverge from the central axis, and often fork before reaching the margin. The true termination is not preserved. It has some resemblance to *Aphlebia patens* Germ., Stein. Fl. v. Wet. u. Löbj. The epidermis of the plant has a smooth aspect marked with the strong impressions of the ribs.

Habitat.—Roof-shales of the Waynesburg Coal, Cassville, W. Va.

7 PP.

FRUITS.

RHABDOCARPUS, Goepp. et Berg.

Rhabdocarpus oblongatus, Sp. nov., Pl. XXXVII, Figs. 8 and 9.

In figs. 8 and 9, Pl. XXXVII, are depicted fragments of a fruit which seems to belong to *Rhabdocarpus* of Goep. and Berg. It is elliptical or oblong in form, and shows 6 or 7 longitudinal ridges. Fig. 8 shows a nut with the pericarp detached, in which the body of the nut seems quite smooth, and marked only by longitudinal lines.

Habitat.—Roof-shales of the Waynesburg Coal, Cassville, W. Va.

———

CARPOLITHES, Sternb.

At only one locality in the Upper Carboniferous strata have we ever seen any nutlets. This is at Cassville, W. Va., where we have found so many and varied plants. The most of the nutlets here come associated with the remains of *Caulopteris.*

Carpolithes bi-carpa. Sp. nov., Pl. XXXVII, Figs. 6 and 7.

(Fruits borne in pairs on a common pedicel, rather rough, oval in shape, with the larger extremity free and tapering to the point of attachment.)

This fruit is evidently somewhat closely allied to *C. fasciculatus,* Lesqx., Vol. II Ill. Rep., Pl. 46, Fig. 7. We have seen four specimens of the fruit, and in all cases they showed the form here figured, i. e., in pairs.

Habitat.—Roof-shales of the Waynesburg Coal, Cassville, W. Va.

Carpolithes marginatus. Sp. nov., Pl. XXXVII, Fig. 1.

(Fruit with a very regular elliptical outline, and margined all around by a raised rim or border. The surface is rather smooth and shows no point of attachment. This fruit was

evidently not very solid or woody in texture, for it left
only a flat leaf-like scale on the shale.)

Habitat.—Roof-shales of the Waynesburg Coal, Cassville, W. Va.

GUILIELMITES, Geinitz.

Guilielmites orbicularis. Sp. nov.. Pl. XXXVII, Fig. 2.

Fig. 2, Pl. XXXVII, represents an impression of a form
which we find in considerable numbers at Cassville. It
agrees so well with the fruit styled *Guilielmites* by Geinitz.
that we place it in that genus. The point of attachment
sometimes shows imperfect marks of a stem, and is shown
by the place from which the lines radiate. It is in nearly all
the specimens excentrically placed, and the woody lines, resembling coarse veins, which radiate from it, fork frequently.
in an irregular manner, as they pass to the margin. It is
without doubt a vegetable impression, since it leaves a film
of coal on the shale which is sharply defined, and cannot
possibly be caused by any compression of the shale, as Car-
ruthers thinks is the case with Geinitz's forms.

Geinitz thought that this fruit was allied to the Palms.
while Schimper considered them to represent the Cycas.
We have no data that can decide this question. The forms
are all orbicular in outline, but vary considerably in size,
that drawn being of average size. Some are considerably
larger.

Habitat.—Roof-shales of the Waynesburg Coal, Cassville, West Virginia.

CONIFERS.

SAPORTÆA, gen. nov.

Leaves simple, subreniform-flabellate. or suborbicular-
cuneate in outline, bordered at the base with a woody rim.
which is apparently an extension of the leaf-stalk ; terminal margin of the leaves, incised more or less deeply ; petiole

long, slender and grooved on the upper surface; nerves departing flabellately from the summit of the petiole and from the woody basal rim throughout its length, under a more or less acute angle, all passing into the lamina, forking sparingly, usually first near the point of insertion, and again once or twice, the branches departing very slightly from each other, and continuing to the terminal margin, nearly parallel to each other, strongly marked, and not closely placed; leaf-substance rather thin, and apparently rather easily torn into strips.

This very interesting plant has no affinity with any fossil form found in the Coal Measures, unless it be allied to Dawson's *Noeggerathia dispar*, "Acadian Geology," Fig. 73.

The plant has characters in common with certain forms of ferns, and also with the coniferous genus Salisburia. The ferns which most resemble this plant are those forms of Adiantum, which like Adiantum reniforme, L. have a flabellate nervation, with a simple frond, marked by a basal nerve which on each side follows the lower border some little distance from the rachis, and then dissolves into branches. This basal nerve however is simply a somewhat more largely developed and freely branching nerve-bundle, and does not differ in function from the adjoining nerve-bundles which pass into the leaf. The forking of the nerves in these ferns is much more frequent than in the fossil, and the nerves or branches are much stronger.

The points of resemblance to Salisburia, on the other hand, possessed by the fossil, are so numerous and striking that Count Saporta, who saw a figure of the specimen depicted in Fig. 1, Pl. XXXVIII, was strongly inclined to consider it a true Salisburia. We think however that if the celebrated French palæobotanist had seen all the figures illustrating this genus, he would not have come to this conclusion.

The following are some of the more prominent features possessed in common by our plant and by Salisburia. They induce us to consider the plant as a new genus of conifers, nearly allied to Salisburia.

Both have the same rather thin leaf substance, with an

incised terminal margin, a grooved petiole, strongly defined, sparingly forked nerves, with branches nearly parallel, and a dichotomous mode of forking which is very characteristic.

The points of difference which induce us to separate the plant generically from Salisburia are the following:—

In Salisburia, the basal cord is merely a branching nerve of no more value than its neighbors. This may be plainly seen on the lower surface of the leaf. The woody bundles in the petiole of the leaf divide at the base of the lamina into two principal nerve-bundles, and each of them on entering the leaf divides into two principal nerves on each side. These by successive forkings, in a dichotomous manner, fill the entire leaf with their branches. The nerve which follows the margin of the leaf has none of the characters of a petiole, and does not send out independent nerves, but simply splits up by dichotomy into a succession of branches of equal value, and which all pursue the same general direction with the principal nerve. The character of the forking is the same with that shown in the principal nerves which enter the leaf more towards the center.

The case is different with the woody border on the fossil plant. This seems to perform the functions of a petiole, or of the rachis in ferns. The mode of departure of the nerves sent off by it, as shown by Fig. 1a is much like that of the lateral nerves from the rachis of a Tæniopteris. It sends off nerves independent of each other, and not mere branches, produced by the splitting up of a parent nerve. We do not find the branches which enter the lamina, in the fossil leaf, to follow so closely the direction of the marginal woody cord, as do the branches in Salisburia. They even, as shown in Fig. 4, attain a direction at right angles with it.

We name the genus in honor of Count Saporta, the celebrated palæobotanist of France.

Saportæa grandifolia. Sp. nov., Pl. XXXVIII, Fig. 4.

(Leaf, with a strong woody cord passing around the base, and descending into a rather slender, long petiole, which is grooved on the upper side. Shape of leaf probably sub-reniform-flabellate; nerves, arising from the summit of the petiole and from the basal rim, the latter strongly diverging from the rim, and soon passing in a direction at right angles to it, on to the terminal margin of the leaf, forking near the point of insertion, and again forking once or twice, the branches diverging but slightly, and soon becoming nearly parallel to each other.)

This fine leaf was seen only in a fragmentary condition. It must have had a considerable expanse. The length of the lamina of the leaf seen is 8 cm.; the width, 9½ cm.; length of petiole seen, nearly 10 cm.; thickness, 6 mm. The furrow on the petiole is very distinct, and the thickness of the basal woody rim is 2½ mm. Fragments of the lamina, seen on the shale containing the portion depicted in Fig. 4, Pl. XXXVIII, show by their position that they belonged to the same specimen, and indicate a leaf at least 15 cm. from base to summit, with a lateral expanse of 20 cm.

Habitat.—Roof shales of the Waynesburg Coal, Cassville, W. Va.

Saportæa salisburioides. Sp. nov., Pl. XXXVIII, Figs. 1–3.

(Leaf, suborbicular-cuneate, flabellate, margin slightly incised into ribbon-like laciniæ; petiole, slender; nerves, arising from the summit of the petiole and from the basal rim, the latter departing under an acute angle, much as in Salisburia, all forking sparingly with the characters of the genus; basal woody cord comparatively slight, but well defined.)

The dimensions of the most perfect specimen seen showed the length from base to summit to be 7½ cm., and the lateral dimensions to be about 10 cm. The plant was evidently much smaller than Saportæa grandifolia, and does not seem to have been so liable to split up, during growth, into laciniæ. The right hand segment of the leaf depicted in Fig. 3, Pl. XXXVIII, shows incisions which seem normal to the species, and not the result of accidents in growth. This plant, also, was always found in a fragmentary condition, though enough of the leaf is preserved in some specimens to show pretty clearly what must have been its shape. Fig. 11, Pl. II, represents a small fragment of a plant which may be different from S. salisburioides, as its texture is thinner, and the fragments found associated with the pieces depicted indicate a leaf of larger size. The fragments are too small to give us any indication of the shape of the entire leaf. They may belong to S. salisburioides in a more advanced stage of growth of the leaf. Plate XXXVIII, Fig. 2, represents a larger fragment, to which also the above remarks may apply.

Habitat.—Roof shales of the Waynesburg Coal, Cassville, W. Va.

Baiera, (Fr. Braun,) emend. Heer.

We follow Heer, in his emendation of the generic character of Baiera, in which he separates it from Salisburia (Cyclopteris,) and unites it with Jeanpaulia.

Baiera Virginiana. Sp. nov., Pl. XXXVII, Figs. 11, 12.

(Leaf, flabellate, divided into numerous laciniæ towards the summit, and narrowing into a wedge shape towards the base, undivided for some distance above the base; laciniæ, slowly diverging, and each forking dichotomously once or twice, divisions strap-shaped and truncate; leaf-substance, thick and leathery; nerves, several in each lacinia, strongly marked, forking once or twice, and proceeding parallel to each other.)

The plant has never been seen entire. Fig. 12 represents the most perfect specimen; Fig. 11 gives a fragment showing more numerous and delicate laciniæ. This plant, in its robust character and thick leaf substance, has much resemblance to Baiera longifolia (Jeanpaulia) Heer, of the Jurassic, given in Vol. IV, Foss. Flor. Arct. Pl. IX, Figs. 1–11. It is very nearly allied to *B. digitata,* Heer, figured by Geinitz in his Dyas, Pl. XXVI, Fig. 2, under the name of *Zonarites digitatus,* Brongt.

The finding of this plant in our Upper Barrens, although it is not specifically identical with the Permian species, is very significant.

GERABLATTINA BALTEATA. PL. XXXVIii, Fig. 5.

This species of cock-roach is represented by the larger part of an upper wing, with its neuration well preserved.

The genus in which it is placed is characterized in a paper on palæozoic cock-roaches, now publishing in the " Memoirs of the Boston Society of Natural History." It is closely allied to Blattina proper, (or Etoblattina, as it must be called,) and next to it, it is of all the fossil genera the richest in species ; and while these belong mostly to the Old World, two of them, including the present form, come from America. *Gerabl. balteata* is distinguished from its neighbors not only by peculiarities in its neuration, and particularly in the course of the internomedian vein and its forked branches, but also by a characteristic which has suggested the specific name, and which does not appear to exist in any other fossil cock-roach, viz: the banded appearance of all the veins and their branches, each being accompanied on either side by a broad, regular border of black carbonaceous matter, upon which are impressed frequent and slight transverse lines. These lines are common to many fossil cock-roaches, but here, instead of traversing the interspaces, as usual, from vein to vein, they do not pass beyond the limits of the black bands. The specimen was found in the roof shales of the Waynesburg coal, at Cassville, W. Va.

S. H. S.

CHAPTER 3.

Summary of Chapter 2, with some conclusions to be drawn from the same.

We come now to the important inquiry as to the relative age of that class of rocks which figure, in the present American nomenclature, under the name of "Upper Barren Coal Measures," since all the plants that have been described in Chapter 2 occur in this series.

In order to have the evidence all before us, we shall arrange these plants in three columns, putting in the first all the plants which we have found in the Upper Barrens; in the second, all of their number which have been reported from the Upper or Lower Coal Measures of the U. S.; and in the third, those which are common to our Upper Barrens and the Permian of Europe.

Arranging the plants in the manner indicated, we get the following:

Table of Distribution of Species.

	U. Barrens of W. Va. and Penn.	Coal Meas. of the U. S.	Permian, Europe.
Equisetides rugosus,	†		†
" elongatus,	†		
" striatus,	†		
Calamites suckowii,		†	†
Nematophyllum angustum,	†		
Sphenophyllum latifolium,	†		
" filiculmis,		†	
" densifoliatum,	†		
" tenuifolium,	†		
" longifolium,	†		†
" oblongifolium,	†		
Annularia carinata,	†		†
" longifolia,	†	†	†

(105 PP.)

Annularia sphenophylloides,			†			†	†
" radiata,							†
" minuta,			†				†
Sphenopteris acrocarpa,			†				allied to *S. oxydata* and *S. lyratifolia*
" coriacea,							
" dentata,			†				
" auriculata,							
" minutisecta,							
" foliosa,							
" Lescuriana,							
" pachynervis,							
" hastata,							
Neuropteris hirsuta,							
" flexuosa,						†	†
" flexuosa longifolia,							
" dictyopteroides,							
" auriculata,					†		†
" odontopteroides,							
" fimbriata,					†		
" cordata,					†		†
Odontopteris obtusiloba,							
" nervosa,							
" pachyderma,							
" densifolia,							
Callipteris conferta,							†
Callipteridium Dawsonianum,							
" oblongifolium,							
" grandifolium,							
" odontopteroides,							
" unitum,							
Pecopteris arborescens,					†		†
" arborescens integri-pinna,							
" Candolleana,				†			†
" elliptica,				†?			
" oreopteridia,				†			
" pennaeformis latifolia,							†
" Miltoni,							†
" dentata,							†
" pteroides,							†
" Pluckeneti,							†
" Pluckeneti constricta,							
" notata,					†		
" Germari,						†	†
" Germari crassinervis,							
" Germari cuspidata,							
" sub-falcata,							
" rarinervis,							
" imbricata,							
" asplenoides,							
" rotundifolia,							
" platynervis,							
" rotundiloba,							
" Schimperiana,							
" pachypteroides,							
" angusti-pinna,							
" Heeriana,							
" tenuinervis,							
" merianopteroides,							
" ovoides,							
" lanceolata,							
" latifolia,							
" inclinata,							
" goniopteroides,							

Goniopteris emarginata,				
" elegans,		?		†
" longifolia,				
" arguta,		† ?		
" elliptica,				
" oblonga,				
" Newberriana,				
Cymoglossa obtusifolia,				
" breviloba,				
" formosa,				
" lobata,				
Alethopteris Virginiana,				
" gigas,				†
Taeniopteris Lescuriana,				Near to *T. multinervis*
" Newberryiana,	†			Near to *T. vittata.*
Rhacophyllum filiciforme,				†
" laciniatum,				
" lactuca,		†		†
" speciocissimum,				
Caulopteris elliptica,				Allied to *C. peltigera.*
" gigantea,	†			Allied to *C. macrodiscus*
Sigillaria approximata,				
" Brardii,				†
Cordaites crassinervis,				
Rhabdocarpus oblongatus,				
Carpolithes bicarpa,				
" marginatus,				
Guilielmites orbicularis,	†			Allied to *G. permianus* Göp.
Saportaea grandifolia,				
" Salisburioides,				
Baiera Virginiana,	†			Allied to *B. digitata.*
Totals,	107	22		28

Before proceeding to an analysis of the table, and the conclusions to be derived from it, we may properly decide what sort of evidence we shall admit to determine the age of a formation, and what is the relative value of that derived from each source. So far as the question of relative value is concerned, this can only arise in the case where we have to consider the conflicting evidence of different classes of organisms, for no one will deny that the life of a period, if well represented, is of the highest value in determining questions of age. For our purpose we need only to consider the claims of the three classes usually most relied upon by Palæontologists, viz: Marine Mollusks, Vertebrates, and Plants. It seems to us erroneous to claim absolutely, that one of these must be valued more highly than another, for the evidence it affords. We must limit the applicability of the evidences from each class.

A priori, we might decide as follows: Mollusks, from the simplicity of their structure, and the nature of the medium in which they live, could not be seriously affected by slight changes of the surrounding conditions, and hence, when not interrupted by cataclysmal agencies, their remains can only be used to denote general changes, requiring long periods of time. They are the hour hand of the palæontological clock. But we must admit the possibility of the existence of special local causes, which may hasten their changes. The same may also occur to modify the normal character of the Vertebrates and Plants. We must however have positive evidence of their existence. Plants, being more dependent on aerial conditions, and less capable of resistance, should give better data for indicating slighter changes, involving shorter periods. They are the minute hand of our clock. Vertebrates are in structure the most complex of the three. They depend in part on plants, and in part, on aerial conditions, or the arrangement of the land and water. Hence they are the most sensitive time indicators, and mark slighter changes requiring shorter intervals. They record the seconds on our clock. Their sensitiveness unfits them for the determination of the longer intervals, which have been founded on the evidences derived from Mollusks or Plants. Relying on them, we would antedate the age of the formation which affords them.

We may hence consider that so far as we can lay down a general rule for the applicability and relative value of the evidence from the three most important classes of organisms, it would be as follows: The evidence of Mollusks should be most weighty in determining long periods; that of Plants, most important in shorter intervals, and that of Vertebrates in the shortest. Of course we must take into consideration all the forms of any one of these classes. It is no more necessary to take a group of plants, in order to get evidence of value, than it is to study the entire collection of marine mollusks, or vertebrates. Modern research shows that many single forms continue to live after the period of their culmination. We must then consider the question of the culmination and decadence of species. The evidence from the existence of exceptionally long lived forms, in any series of strata, must be considered of slight value.

Again, in assigning their relative value to the component parts of any of these three classes of organisms, we must consider that representative or closely allied forms should have hardly less value than identical ones, in certain cases. This is especially true where we find many closely related, and few identical species. We must not necessarily conclude that the age of two formations in such a case is different, but that surrounding conditions had sufficient power to modify specific characters. We must assign considerable value to resemblance, or difference, in type, for a change of type implies a change in the conditions of existence.

It is good evidence that we have to deal with a more recent formation, when we find it to show a decadence of old forms, and an introduction of new ones, destined to reach their culmination at a later period. Thus if we find, in a series of rocks, plants characteristic of the Carboniferous formation, and perceive that these die out and disappear, we should not conclude from their mere presence, that the age of the strata is Carboniferous, but rather that it is Permian. So also the finding of genera and species, even identical with those of the Trias or Jurassic, would not necessarily imply a Triassic or Jurassic age. If we find them to be exceedingly rare, their presence is rather indicative of a formation older than the Trias or the Jurassic.

It is only by taking into consideration all the above named characters, and other points which may be presented by the entire body of specimens, that we can determine the nature of the evidence offered by the life of a formation. It will not suffice to say arbitrarily, that this or that feature is without value as evidence. Circumstances might reverse the normal relative weight of the evidence from the

several sources and give preponderating weight to what
would, if unaffected by them, have slight value.

Having thus established the evidence of one class of organisms, we must combine it with that of any other class
afforded by the strata, and the general facies of the entire
life only can be used in determining the age. But the
evidence from this source must not stand alone, if we can
supplement it with that derived elsewhere, we must search
all possible sources.

A source from which we may often derive evidence of
great value is the stratigraphy, and especially the lithology.
Many geologists unduly depreciate the value of the latter.
It is easy to see that where the strata have such a character,
that they could only be formed under special conditions,
they must have a certain value as evidence, especially when
this is combined with the diminution, or disappearance of
beds characteristic of a certain formation. Thus in ascending from a known carboniferous horizon, to superimposed
formations, if we find the coal abundant in the lower beds,
and disappearing in the upper, while great masses of limestone and fine grained red shales come in, surely this would
be weighty evidence to show that carboniferous conditions
had changed to Permian. If the life of the period is very
scantily represented by fossil forms, cases might occur
where we would be called upon to determine the relative
value of the two, and we would find the evidence of Lithology of superior weight. It is not impossible to find cases
where the evidence of Lithology resembles in character
and degree that of fossils. Certain strata may have such
a peculiar character, that when their eroded fragments enter
into the composition of later formed brecciæ, or conglomerates, they may be recognized with certainty a hundred
miles and more from the parent source. This is actually the
case with certain conglomerates of the eastern portion of
Virginia, which are of Potsdam quartzites.

We may also employ the evidence to be derived from the
" Breaks" and physical changes found to occur. If two
formations are separated by strata giving evidence of a
change in the prevailing conditions sufficient to cut off or

modify the life of the lower, the existence of this convulsion alone would be of weight in indicating a later age for the higher formation. We need not have necessarily unconformity.

Let us now turn to the table of species, and determine the bearing of the evidence to be derived from it on the question now before us.

Considering first the identity of the species named in the table, we see that out of 107 found in the Upper Barrens of West Virginia, 22 occur in the Coal Measures proper, while 28 are found also in the Permian of Europe, according to Göppert, Weiss, Schimper, Geinitz, Grand'Eury, Gutbier, Heer and others.

Of the 22 species which are common to the Upper Barrens and to the Coal Measures below them, 16 are also found in the European Permian, leaving 6 not hitherto found in the Permian. Of these 6, one is given by Bunbury as occurring at Frostburg, Maryland. This is Pecopteris elliptica. Bunbury makes no distinction of horizons at Frostburg, and as the Upper Barrens occur there, it is almost certain that this species should be credited to them. This leaves 5 species, viz: Sphenophyllum filiculmis, Neuropteris hirsuta, N. fimbriata, Pecopteris notata, and Goniopteris arguta. Goniopteris arguta is found by Prof. Lesquereux only in the flora of Illinois, which flora is peculiar in possessing many Permian types. The presence of Neuropteris hirsuta may be explained by the fact that it is a long lived plant, enabled by vigor of growth and constitution to pass above the horizon which it characterizes. Pecopteris notata occurs in the Anthracite Coal Region of Pennsylvania. The geological horizons are not yet fully worked out there, but enough is known of the height of the column of coal measures in the deepest parts of the basins to make it almost certain that the horizon of the Waynesburg Coal occurs there.

But even supposing that these 6 species should be credited without qualification to the true coal measures, the number of true coal measure species in the above list would be surprisingly small.

Grand'Eury, in his account of the Permian of Central France, in his "Flor. Car. du Departement de la Loire, et du Centre de la France," says that the Upper Coal Measures' flora passes insensibly into the Permian, there being a mixture of the two floras, and that he often finds it almost impossible to draw the line of separation. He states that the researches of himself, Weiss, and Göeppert, have raised the number of species common to the Coal Measures and to the Permian to fifty.

Of these 6 species, Neuropteris hirsuta is the only one found above the Waynesburg Sandstone, so that whatever significance their presence in the transition beds between the Waynesburg Coal and Sandstone may have, this is lost in passing above them.

Let us now consider the species common to the Upper Barrens and to the European Permian. Of these 28 species, 12 have never been found in the Coal Measures of the United States, and two, *Callipteris conferta and Alethopteris gigas*, are exclusively Permian. The presence of Callipteris conferta, is usually considered as proof of the Permian age of the strata containing it. Odontopteris obtusiloba, though commencing in the highest strata of the Carboniferous, as Grand'Eury shows, is a characteristic Permian plant. Annularia carinata, if distinct from Annularia calamitoides, would be peculiarly a Permian plant. It seems to us, however, to be the same with A. calamitoides.

Passing to representative and allied species, we have some whose presence bears weightily in the argument.

Baiera Virginiana differs from B. digitata, the Permian plant, chiefly in its greater size and robustness. The genus Baiera begins in the European Permian.

Taeniopteris Lescuriana is the representative of T. multinervis, an exclusively Permian plant: while T. Newberryiana is closely allied to T. coriacea, also Permian. Both, in many features, seem prototypes of much more recent forms found in the Mesozoic.

Sphenopteris coriacea is closely allied to the Permian

species S. oxydata, and is its representative in our flora. It is also allied to the S. lyratifolia of Heer.

Grand'Eury is disposed to regard the Sphenopterids of this type as forms of Callipteris. This view is confirmed by the leather-like character of S. coriacea, and by the fact that it occurs associated with Callipteris conferta. At any rate the type of Sphenopteris shown in S. coriacea, is exclusively Permian.

The genus Cymoglossa, founded by Schimper on one species, is exclusively Permian. Its very considerable development in our flora is of great weight as indicating a period later than the true Coal Measures. The plants of this genus are evidently modifications of the Goniopteris type, which is itself characteristic of the closing period of the Coal Measures and of the Permian.

The genus of Nematophyllum, in the absence of mid-nerve in the leaflets, in their great elongation without marked change of width, and in their union, at least at base, is allied to the genus Schizoneura, which begins in, and is highly characteristic of the Permian.

We may state here that we hesitated for sometime about separating this plant from Schizoneura, and were finally induced to do so from the fact that we nowhere saw the leaflets united together, and attached unmistakably to a stem. The union of the leaflets in the young state is the most important character of the genus Schizoneura. Yet we saw many fragments of leaves, having precisely the texture, striation, &c. of the leaflets of Nematophyllum, which were an inch or more wide, and showed a splitting, to a greater or less depth, into thread-like laciniæ of the width of the leaflets of Nematophyllum. These were never attached, and if they represent the united younger leaves of a Schizoneura they must be very deciduous.

The authors of the European species of Schizoneura, however, seem to attach little value to this union in the younger leaves. Schizoneura Meriani, Schimp. of the European Trias, closely resembles our Nematophyllum in many other features, as well as in the separation of the leaves. Heer in his "Pfl. der. Trias u. Jura," states

8 PP.

that the union of the leaves has never been observed in this plant, and yet he admits it as a Schizoneura, although with doubt.

Among the fruits we find Guilielmites orbicularis, closely allied to the Permian species of Geinitz, G. permianus.

The decadence in the Upper Barrens of certain plants highly characteristic of the Coal Measures proper is another feature pointing strongly to their age as Permian. This feature, as is well known, characterizes the Permian of Europe, and is of hardly less value than the identity of species in distinguishing this formation.

The European Permian, according to Grand 'Eury, possesses the last representatives of *Lepidodendron, of Sigillaria*, and of *Calamites*; while it contains many *Pecopterids*, the greater portion of them have become subarborescent. The *Alethopterids* are rare, as are also the *Odontopterids*, which have the Mixoneura type of nervation in this formation. The *Callipterids* now make their first appearance.

All these conditions are fulfilled, in the most striking manner, in our Upper Barrens.

Not a single *Lepidodendron* occurs. Only two species of *Sigillaria* are found. One of them, S. Brardii, passes up into the Permian, and the other is of the peculiar type of S. Brardii, which is more characteristic of the Permian than of the Carboniferous. Only one *Calamite* occurs, and this also passes up into the Permian. Of the *Alethopterids* we get only two species, one, A. Virginiana, more nearly allied to Callipteridium than to the Alethopterids which characterize the Coal Measures proper; the other, A. gigas, is a characteristic Permian form.

Of *Odontopteris* we find 4 species only, all with Mixoneura nervation, and one, O. obtusiloba, rather Permian than Carboniferous.

Nearly all the *Pecopterids* show the arborescent character. This is seen in the greater expanse of their fronds, and in the considerable size of their stipes, many of which are from four to six inches in diameter. Most of the species, also, belong to Schimper's section Cyatheides, of which Pecopteris arborescens is the type.

While the marked decadence of characteristic Carboniferous forms has affected the facies of the flora of the Upper Barrens, a still greater change is produced by the introduction of new features, among which we find the first appearance of types destined to reach their culmination in the Mesozoic. We will specify only a few of these new features.

The *Neuropterids* show a Permian character in the tendency of the middle nerve to split up, and in the approach of their nervation to the Mixoneura type of the Odontopterids. In this feature, and in the great size of the rachis, they resemble the Permian Neuropterids of the type of N. Dufresnoyii Brongt. The *Sphenopterids*, in the delicacy of their foliage, and the character of their lobing, differ much from those of the Carboniferous, and show affinities with Mesozoic forms. Sphenopteris minutisecta resembles a Thyrsopteris; S. acrocarpa, in the foliage of the sterile plant, resembles this genus; while the only fossil plant known to us which has a somewhat similar fructification, is the Acropteris cuneata of Schenk, found in the Rhaetic of Europe. Our Equisitides elongatus, in the long linear divisions of the sheath, consolidated except at the top, and terminating with obtusely rounded ends, as well as in the strong middle nerve which runs down the surface in the middle of each leaflet, is more like the peculiar Equisetum triphyllum, of Heer, from the Trias of Switzerland, than any other described fossil form. Pecopteris merianiopteroides, is strikingly like Heer's Triassic genus, Merianiopteris; while Pecopteris pachypteroides, has many of the features of Pachypteris.

A very interesting feature shown in some of the forms of Pecopteris, and Callipteridium, is the foreshadowing of some of the characters of the Mesozoic Pecopteridæ, of the type of Pecopteris Whitbyensis. In the falcate, acute pinnules, the long, almost linear pinnae and the nervation, we have the features of the genus Cladophlebis, as limited by Saporta.

The appearance in the Upper Barrens of Saportaea, a genus so nearly allied to Salisburia or Gingko, is of great importance, both as indicating that great changes were occurring in the flora, and also as establishing the fact that the peculiar coniferous type, which in Gingko or Salisburia, attains such importance in the Jurassic, had already made its appearance in the Permian.

Heer, in his fourth volume of the Arctic Flora, shows that the Gingko, or Salisburia, had acquired a great development in the number of forms and in the abundance of individuals in the Jurassic. Hence *a priori* we should expect to find the first appearance of the type in a much older formation. It is interesting to note also, that the genus Baiera, as limited by Heer, which appears in such development in the Jurassic, associated with Salisburia, makes its first appearance in our Upper Barrens along with Saportaea, which we may consider as the ancestor of the Jurassic forms of Salisburia.

There are two important plant-bearing horizons in the Upper Barrens. The lowest, is the shale which forms the roof of the Waynesburg coal. This contains all the species which have ascended from the Coal Measures proper, along with many new forms. The second horizon is that of the Washington Coal, where we find all the Coal Measure species (with the exception of Neuropteris hirsuta) to have disappeared, and note the first occurrence of Callipteris conferta, Sphenopteris coriacea and others. Hence the evidence of the Permian age of this series of strata, lying above the Waynesburg Sandstone, is not at all weakened by the presence of characteristic Coal Measure forms.

The evidence from animal life is not weighty, but so far as it goes it is in favor of the Permian age of the strata in question. The limestones and shales affording the animal forms found, which are mollusks, and bivalve crustaceans, appear to have been deposited in fresh water, and this accounts for the uncertain character of the evidence. Among them we find species of the *Cypris* and *Estheria* very closely allied to those of the Trias. A univalve mollusk also, of almost microscopic proportions, is very abundant in certain layers.

Not a single *species* of the very abundant and varied molluscan forms of the Coal Measures passes up into the Upper Barrens, and the only *genus* from the lower measures that we have ever seen in the upper, is *Solenomya*, which is represented by a form quite close to *S. permiensis*.

A suite of specimens, representing about all the animal life that we find in the Upper Barrens, was submitted to Prof. James Hall, the eminent paleontologist of Albany, N. Y., and he gave it as his opinion that there was nothing among them which might not be of Permian age.

We may next inquire whether we have evidence of any considerable change which would suffice to produce an important effect, and alter the conditions which prevailed in the lower beds, which all recognize as of Carboniferous age. For this purpose we must turn to the general geology of the district. From this we find, after ascending above the Pittsburg Coal, and its associated coals the Redstone and Sewickley, two horizons which give evidence of extensive physical changes.

The first of these horizons marks the general submergence which produced the important limestones and calcareous shales which occupy much of the interval between the Sewickley and the Waynesburg. We find no plants until we reach the roof shales of the last named coal. These shales, as we see from our analysis of the table, contain nearly all the characteristic Carboniferous plants which pass into the Upper Barrens, mixed with a great number of new forms. The physical change here was not sufficient to entirely alter the flora.

The second horizon of changing conditions, is found in, and immediately above the Waynesburg Coal. In the rapid fluctuations in thickness of the clay parting of this coal we see the first indications of unquiet, and of the approach of that much greater disturbance which produced the important Waynesburg Sandstone which in its extent and character gives ample evidence of wide spread change.

The Waynesburg Sandstone often rivals the great Conglomeratic Sandstone, which forms the base of the Productive Coal Measures in the amount of pebbles which it contains. It is often 75 feet thick, and in expanse is co-extensive with the Upper Barrens. To form an idea however of the amount of the change required to produce this great mass, we must not simply consider the character of the stratum *per se*, but must contrast it with the strata which immediately precede it. Leaving out of view the Waynesburg Coal, all the rocks for a considerable distance under it are either limestones or fine grained shales, which show that the deposition of sediment must have taken place under conditions of general quiet. The shale roof of the Waynesburg Coal is not always present. We sometimes find the sandstone lying immediately on the coal, and even descending into it.

When, then, in such localities we see the immense sandstone loaded with pebbles lying immediately upon the coal with its subjacent fine-grained beds, we are forcibly impressed with the magnitude of the change which has taken place. The character of the pebbles also is significant. They are not of sandstone but of quartz, and hence must have been brought from remote localities.

Let us now consider what is the evidence from the Lithology of the strata of the Upper Barrens. Leaving out of consideration the finding of a conglomerate at the base of the series, a feature which it has in common with the Permian of Europe, we find in it a great deal of red shale, another feature of the Lower Permian of Europe. These red shales occur in beds 20'–30' thick, sometimes commencing immediately above the Waynesburg Sandstone. They are a pretty constant feature, and are often, as at Bellton, several hundred feet thick. These features, taken alone, are not entitled to much weight, except as showing conditions unfavorable for the formation of coal, as they are found also in the barren portions of the Carboniferous formation proper. Besides these characteristics which mark the Lower Permian of Europe, the Upper Barrens have some in common with the Zechstein or Upper Permian, in the presence of a large amount of limestone.

It is a significant feature that these limestones are devoid of marine fossils, showing that the sea had access at no time during their formation.

The evidence from the total disappearance of coal beds

●

in the higher portions of Upper Barrens, and from the extremely small amount of it found in the lower portions, is of more value, as indicating a great change from the conditions which prevailed during the Carboniferous proper. The beds of coal gradually disappear as we pass upwards, and with the exception of the Washington Coal, are never more than one or two feet thick, while the uppermost 200 or 300 feet contain none at all. This diminution of the coal is accompanied with a great loss in the amount of plant life. Only about 20 p. c. of the forms existing below the Waynesburg Sandstone pass above it, and of these, many are sparingly represented, and seem to be in process of extinction. These features are represented to be characteristic of the European Permian. Grand'Eury, in his Fl. Car. du Dep. de la Loire et du Centre de la France," states that he finds the Permian to be marked by a diminution of coal, and a decadence of the flora. This is what we would expect *a priori*, if we should regard the Permian, not as a distinct formation, but as the close of the Carboniferous. The idea of its distinctness arose from the fact that the Permian was first studied in Saxony and other countries where a complete physical break exists, and where the evidence of gradual passage could not be derived from the stratigraphy and fossils. More extended study of the formation in such countries as France shows that this break is not universal, and that the passage from the Carboniferous proper to the Permian is a gradual one. The investigations of Weiss, Grand'Eury, and others, indicate that the Permian is merely the closing period of the Carboniferous.

In the United States, there is no unconformity in the strata from the lowest beds of the Carboniferous to the highest stratum found in the Appalachian Coal Field. In view of this, it is remarkable that we should find such great changes in the flora as we actually do discover.

To sum up finally the evidence derived from all sources, we find ourselves irresistibly impelled to the conclusion, that the age of the Upper Barrens of the Appalachian Coal Field are of Permian age, by a consideration of:

1. The Evidence from the Identity of Species.
2. The Evidence from Allied Species.
3. The Evidence from the decadence of Coal Measure forms.
4. The introduction of Types characteristic of later Formations.
5. The existence of an important Physical change at the beginning of the Series.
6. The nature of the Lithology ; the disappearance of Coal ; the diminution in the Amount of Plant Life.

The evidence of the animal life of the Upper Barrens is of no particular weight in determining the question. So far as it goes, it is favorable to the conclusion that the age is Permian.

It might perhaps be best to separate the roof shales of the Waynesburg Coal and Waynesburg Sandstone from the beds overlying the sandstone, and as they contain a mixed flora, consider them as transition beds of Permo-Carboniferous age. Perhaps the strata down to and including the great limestone overlying the Sewickley Coal should be included with these, but in the absence of fossils this cannot be decided. The beds above the Waynesburg Sandstone should, however, be considered as strictly Permian.

If this conclusion be correct, it will have an important bearing on the history of the changes which have affected the Physical Geography of our portion of the North American Continent. Our great Appalachian Revolution would have occurred at the close of the Permian Period, and instead of standing almost alone, would be in harmony with those mighty changes which elsewhere operated at the close of the Permian to extinguish the forms of Palæozoic life.

It would also explain the absence of Permian beds in the Mesozoic areas of the eastern portion of the Continent, and the Triassic age of the oldest beds there found. For, if our views be correct, the basins in which these beds were laid down were formed at the close of the Permian, instead of the Carboniferous proper.

DESCRIPTION OF PLATES.

PLATE I.

Figs. 1–4. Equisetides elongatus, Spec. nov.
"   1. Equisetides elongatus. A large fragment.
"   1^a. Enlarged rib of same to show mid-nerve.
"   2. Equisetides elongatus, a smaller and more slender specimen.
"   3. The same. A specimen showing what is probably the base of the sheath.
"   4. The same. A specimen showing the insertion of sheath.
"   5. Equisetides striatus. Spec. nov.
"   6. Equisetides rugosus, Schimp.
"   7. Sphenophyllum densifoliatum, Spec. nov.
"   7^a. A pair of leaflets of the same enlarged.
"   8. Sphenophyllum filiculmis, Lesqx.
"   8^a. Leaflet of the same enlarged.
"   9. Sphenophyllum tenuifolium, Spec. nov.
"   9^a. Leaflet of the same enlarged.
" 10–11. Sphenophyllum latifolium, Spec. nov.
"   10. Sphenophyllum latifolium. Specimen showing the toothed and irregular terminal border.
"   10^a. A leaflet of the same enlarged.
"   11. The same, with the terminations of the leaflets wanting.

PLATE II.

Figs. 1–5. Nematophyllum angustum, Gen. nov. et spec. nov.

(121 PP.)

Figs. 1. Nematophyllum angustum. A fragment of stem
with several joints.
" 2. Two stems of the same apparently diverging
from a point of junction.
" 3. A fragment of the same showing a whorl of
leaves united at base into a ring.
" 4. A fragment of the same showing apparent
union at the base of the leaflets.
" 4ᵃ. Enlarged leaflet of the same to show the
striations.
" 5. A fragment of the same showing more or
less union at the base of the leaflets.
" 6. An undetermined specimen. Apparently
it is a portion of a flabellate leaf.

PLATE III.

Figs. 1-3. Sphenopteris acrocarpa, Spec. nov.
" 1. Sphenopteris acrocarpa, sterile plant.
" 1ᵃ, 1ᵇ. Enlarged pinnules to show nerves.
" 2, 3. Portions of fertile plant.
" 2ᵃ. Fertile pinnules of the same enlarged.
" 2ᵇ. Fructification of the same still more enlarged to show group of sori.

PLATE IV.

Figs. 1-5. Sphenopteris acrocarpa.
" 1. Sphenopteris acrocarpa. Portion of sterile frond
towards the summit.
" 1ᵃ. Enlarged pinnule of the same to show the
nerves and segments.
" 2. Portion of the same plant from near the extremity of a compound pinna.
" 3. Portion of the plant from near the summit
of the frond.
" 4. Portion of the same from towards the base
of the frond.
" 5. Portion of the plant from near the extremity
of a compound pinna.
" 5ᵃ. Pinnule of the same enlarged to show nervation.

Plate V.

Figs. 1–4. Sphenopteris minutisecta. Spec. nov.

"   1. Sphenopteris minutisecta.   A portion of the frond from towards the base.

"   1ᵃ. A pinna of the same enlarged.

"   2. A portion of the same plant from higher up in the frond.

"   3. A portion of the same plant from the summit of a compound pinna.

"   4. A portion of the same from near the summit of the frond.

"   4ᵃ. A pinnule of the same enlarged.

"   5. Sphenopteris coriacea. Spec. nov.

"   5ᵃ. Pinnules of the same enlarged.

"   6. A fragment of a pinna.

"   7. Sphenopteris dentata, Spec. nov.

"   7ᵃ. Pinnules of the same enlarged.

"   8. A fragment of the same from lower down in the frond.

"   9. Sphenopteris foliosa, Spec nov. A pinna.

"   9ᵃ. A pinna of the same enlarged.

"   10. Fragment of an ultimate pinna from a lower part of the frond.

"   11. A fragment from near the summit of a compound pinna of the same plant.

Plate VI.

Fig.   1. Sphenopteris Lescuriana, Spec. nov.

"   1. Sphenopteris Lescuriana. A compound pinna.

"   1ᵃ. A portion of the extremity of one of the ultimate pinnæ of the same enlarged.

"   1ᵇ. Pinnules of the same enlarged.

Plate VII.

Fig.   1. Summit of the compound pinna given in Fig. 1, Plate VI. (The size of the plate did not permit the insertion of the entire figure on one plate.)

Fig.   2. Terminal portion of a compound pinna corresponding to Fig. 1, Plate VI.

"   3. Sphenopteris auriculata, Spec. nov.   A portion from the upper part of the frond.

" 3^a Enlarged pinnules from different parts of 3^b the same. Fig. 3^c is a portion of a pin- 3^c nule from the lower part of the specimen.

"   4. A portion of the lower part of the frond of the same plant.

" 4^a. An enlarged portion of an ultimate pinna or pinnule of the same.

"   5. Sphenopteris pachynervis, Spec. nov.

" 5^a. An enlarged pinnule of the same.

"   6. Summit of a pinna of the same.

"   7. Sphenopteris hastata, Spec. nov.   A portion of a pinna.

" 7^a. An enlarged pinnule of the same.

PLATE VIII.

Fig. 1. Neuropteris flexuosa, Brongt., var. longifolia.

"   2. Neuropteris platynervis, Spec. nov.

"   3. Neuropteris dictyopteroides, Spec. nov. An entire pinnule.

"   4. The same plant. Fig. 4^a is an enlarged portion of the base of one of the pinnules of the last, to show the reticulation.

"   5. The same plant. The specimen shows the summit of the pinna or frond.

"   6. Neuropteris flexuosa, Brongt.   A small abnormal form.

"   7, 8. Neuropteris hirsuta. Lesqx.   Fructified leaflets.

" 8^a. A portion of the pinnule enlarged, to show sori on the veins.

PLATE IX.

Fig. 1. Neuropteris odontopteroides. Spec. nov.   A pinna pinna from the lower part of the frond.

" 1^a. An enlarged pinnule of the same.

Fig.   2, 3, 4, 5, Pinnæ of the same, showing vari-
                   ations in shape and distance of pinnules.
"   2ᵃ. Enlarged pinnule of Fig. 2.
"   6. Summit of pinna of the same plant.

PLATE X.

Figs. 1-2. Odontopteris nervosa, Sp. nov.   The specimen
                   shows the insertion of the pinnules.
"   2. Terminal portion of a pinna of the same.
"   3. Odontopteris densifolia, Spec. nov.
"   3ᵃ. Pinnule of the same, showing nervation.
"   4. Odontopteris obtusiloba. Naum. Variety. rari-
                   nervis.
"   4ᵃ. Pinnule of the same, showing nervation.
"   5-10. Odontopteris pachyderma, Spec. nov.
"   5. An entire pinna.
"   6. The same.   A portion of a pinna from lower
                   down in the frond, showing crenulated
                   pinnules.
"   7. A fragment of the same, showing the most
                   common form of pinnules.
"   7ᵃ. Enlarged pinnule of the same.
"   8. A fragment of the same, showing a portion
                   of the upper part of the frond.
"   9. A portion of the same, showing insertion
                   of the pinna, and the auriculate pinnules
                   at their base.
"   10. Basal portion of a large pinna of the same.
"   10ᵃ. Basal leaflet of the same, enlarged.
"   11. Neuropteris, species not determined. Basal por-
                   tion of a large rachial leaflet.

PLATE XI.

Figs. 1-4. Callipteris conferta, Brongt. The normal form.
"   1ᵃ. Enlarged pinnules of the same.
"   2. Terminal portion of a pinna.
"   3. Fragment from the upper part of a com-
                   pound pinna.

Fig.   4. Fragment of a compound pinna from the
                  upper part of the frond.
"   4ᵃ. Enlarged ultimate pinna of the same.
"   5, 6, 7. An undetermined plant, probably a Callipteri-
                  dium, 5ᵃ gives the nervation of a pinnule
                  of Fig. 5.

PLATE XII.

Figs. 1–5. Callipteridium oblongifolum, Spec. nov.
"   1. Shows a portion of the lower part of the
                  frond.
"   1ᵃ. A normal pinnule of the same enlarged.
"   1ᵇ. A portion of the basal pinnules in the lower
                  pinnæ of the specimen enlarged to show
                  the grouping of the lateral nerves.
"   2. Fructified form of the same plant.
"   2ᵃ. Pinnules of the same enlarged.
"   3. Fragment of a pinna with large pinnules.
"   4. Shows another form of the same plant.
"   4ᵃ. An enlarged pinnule of the same.
"   5. A fragment from near the summit of the
                  frond.

PLATE XIII.

Figs. 1–2. Callipteridium Dawsonianum, Spec. nov.
"   1. A portion of a compound pinna.
"   1ᵃ. Enlarged pinnules from the lower pinnæ of
                  the same.
"   1ᵃ. Enlarged pinnules from the upper pinnæ of
                  the same.
"   2. A portion of an ultimate pinna from a lower
                  part of the frond.

PLATE XIV.

Figs.   1. Callipteridium Dawsonianum. The terminal por-
                  tion of the compound pinna shown in
                  Pl. XIII, Fig. 1.
"   1ᵃ. A portion of an ultimate pinna of the same
                  enlarged.

Figs. 2, 3. Callipteridium unitum. Spec. nov.
"   2. Represents a part of the frond higher than
               that shown in Fig. 3.
"   2ᵃ. Enlarged pinnules of Fig. 2.
"   3ᵃ. Enlarged pinnules of Fig. 3 to show nerva-
               tion and constricted base of pinnules.

PLATE XV.

Figs. 1–4. Callipteridium grandifolium, Spec. nov.
"   1. Shows the irregularly lobed pinnules of the
               lower part of the frond.
"   2. Gives the normal pinnules.
"   2ᵃ. An enlarged pinnule of the same.
"   3. Shows a heteromorphous form with more
               remote pinnules.
"   4. The terminal portion of a compound pinna.
"   4ᵃ. An enlarged pinnule of the same.

PLATE XVI.

Fig.   1. Callipteridium odontopteroides, Spec. nov.
"   1ᵃ. Enlarged pinnules of the same.
"   2–4. Callipteridium grandifolium. Spec. nov.
"   2. The summit of a compound pinna.
"   3. Shows a form with elliptical pinnules.
"   4. A fructified portion of the same.

PLATE XVII.

Fig.   1. Pecopteris elliptica Bunb.
"   1ᵃ. Enlarged pinnule of the same.
"   2. Pecopteris rotundiloba, Spec. nov.
"   2ᵃ. Enlarged pinnule of the same.
"   3. Pecopteris species?
"   3ᵃ. Pinna of the same enlarged.
"   4–5. Pecopteris pennaeformis, Brongt., Var. lati-
               folia.
"   5ᵃ. Pinnule enlarged.
"   6. Pecopteris species? The fragments show very
               large sori.
"   6ᵃ. Enlarged pinnules of the same.

PLATE XVIII.

Figs. 1–6. Pecopteris platynervis, Spec. nov.

" 1. Gives the normal pinnules from the middle part of the frond.

" 2. A compound pinna of the normal form. Fig. 2^a. Pinnules of the same enlarged. Fig. 2^b. Gives a portion of the same pinnules still more enlarged, to show the double character of the lateral nerves.

" 3. A portion of a compound pinna from the upper part of the frond.

" 3^a. An enlarged pinna of the same.

" 4. The summit of a compound pinna.

" 5. A portion from near the summit of a compound pinna.

" 5^a. An enlarged pinna of the same.

" 6. An ultimate pinna from near the base of the frond. Fig. 6^a. Enlarged pinnules of the same with more complex nervation. Fig. 6^b. A fragment of the same still more enlarged to show the flat nerves.

PLATE XIX.

Figs. 1–7. Pecopteris Germari (Weiss). Font. & White.

" 1. A compound pinna from the upper part of the frond.

" 2. A portion of the lower part of the frond.

" 2^a. Enlarged ultimate pinna of the same.

" 3. A small fragment of an ultimate pinna from the lower part of the frond.

" 3^a. A pinnule of the same enlarged.

" 4. The summit of a compound pinna.

" 5. The basal portion of the same.

" 6. A portion from the summit of the frond.

" 6^a. Enlarged pinnules of the same.

" 7. Summit of compound pinna, showing a more distant arrangement of the ultimate pinnæ.

PLATE XX.

Figs. 1–3. Pecopteris, Candolleana. Brongt.
" 1. A portion of a pinna showing fructification.
" 1ᵃ. An enlarged pinnule of the same.
" 2. A portion of a pinna from the lower part of the frond, showing the beginning of the lobing of the pinnules.
" 3. A portion of a pinna, showing more complex nerves than the normal pinnules.
" 3ᵃ. Enlarged pinnule of the same.
" 4. Pecopteris Germari, var. cuspidata. Var. nov.
" 4ᵃ. Pinnules of the same, enlarged.
" 5. Pecopteris Germari, var. crassinervis. Var. nov.
" 5ᵃ. Pinnule of the same, enlarged.
" 6–8. Pecopteris rarinervis, Spec. nov.
" 7, 8. Show portions from near the extremity of compound pinnæ.
" 6ᵃ. Enlarged pinnules of the same.

PLATE XXI.

Figs. 1,2. Pecopteris subfalcata, Spec. nov.
" 1. Gives the normal form.
" 1ᵃ. Enlarged pinnules of the same.
" 2. Gives an abnormal form.
" 3. Pecopteris Pluckeneti. Brongt. Var. constricta.
" 3ᵃ. Enlarged pinnule of the same.
" 4–5. Pecopteris Pluckeneti. Brongt.
" 4ᵃ. Enlarged pinnule of Fig. 4.

PLATE XXII.

Figs. 1–5. Pecopteris dentata, Brongt.
" 1ᵃ. Enlarged pinnules from the lower part of Fig. 1.
" 1ᵇ. Enlarged pinnules from the upper part of Fig. 1.
" 2. A form corresponding with Pecopteris plumosa of Brongt.
" 3, 4, 5. Give different forms of P. dentata.

9 PP.

PLATE XXIII.

Fig. 1. Pecopteris imbricata, Spec. nov.
    "   1^a. Enlarged pinnules of the same.
  " 2-3. Pecopteris Miltoni, Brongt. These are varietal
                 forms of this polymorphous plant.

XXIV.

Figs. 1-5. Pecopteris Schimperiana, Spec. nov.
    "   1. Gives the normal form.
    "   1^a. Enlarged pinnules of the same.
    "   2. A portion from the lower part of the frond.
    "   2^a. Enlarged pinnules of the same.
    "   3, 4. Give another form somewhat different.
    "   5. A portion from the upper part of the frond.
    "   6. Pecopteris rotundifolia. Spec. nov.
    "   6^a. Portion of a pinna of the same enlarged.
    " 7, 7^a. Pecopteris species?

XXV.

Fig. 1. Pecopteris asplenioides, Spec. nov.
    "   1^a. Fertile and sterile pinnules enlarged.
    "   2. Pecopteris goniopteroides, Spec. nov.
    "   2^a. Enlarged pinnules of the same.
  " 3-7. Pecopteris Heeriana, Spec. nov.
    "   3. Fructified portion of the plant.
    "   3^a. Enlarged pinnule of the same.
    "   4. Portion of the sterile plant from the upper
                 part of the frond.
    "   4^a. Pinnules of the same enlarged.
    "   5. Portion of the lower part of the frond show-
                 ing crenulated pinnules.
    "   6. A portion of a pinna from near the summit
                 of the frond.
    "   7. Summit of compound pinna or frond.
    "   7^a. Pinna of the same enlarged.

PLATE XXVI.

Figs. 1–4. Pecopteris pachypteroides, Spec. nov.
             1. Portion of the frond showing incipient teeth in the pinnules of the lower pinnæ.

             1^a. Enlarged lower part of the frond showing the incipient teeth.

       1^b. 1^c. Enlarged pinnæ from the upper and middle portions of the frond.

             4. Summit of a compound pinna.

PLATE XXVII.

Figs. 1–3. Pecopteris angustipinna, Spec. nov.
             1. Normal form.

             1^a. Enlarged pinnules of the same.

             2. Portion of the lower part of frond.

             3. Portion of the upper part of the frond.

             3^a. Enlarged pinnules of the same.

     4. 4^a. Pecopteris species?

     5. 5^a. Pecopteris species?

      6. Pecopteris arborescens?

PLATE XXVIII.

Figs. 1–4. Pecopteris tenuinervis, Spec. nov.
             1. A portion from the lower part of the frond showing undulate pinnules.

             1^a. Enlarged pinnules of the same.

             2. Compound pinna from the middle portion of the frond.

             2^a. Enlarged pinnules of the same.

             3. Portion from the upper part of a compound pinna.

             4. Fructified portion of the plant.

             4^a. Pinnules of the same enlarged.

             4^b, 4^c. Sori enlarged.

XXIX.

Figs. 1–2. Pecopteris merianiopteroides, Spec. nov.
             2. Summit of a pinna.

Figs. 3, 3^a. Pecopteris ovoides, Spec. nov.
" 4, 4^a. Pecopteris inclinata, Spec. nov.
" 5, 6, 6^a. Pecopteris latifolia, Spec. nov.
" 7, 8, 9. Pecopteris lanceolata, Spec. nov.
" 8. Extremity of a pinna.
" 9. A pinna from the lower part of the frond.

PLATE XXX.

Figs. 1^a. Goniopteris elliptica. Spec. nov.
" 2, 2^a. Goniopteris Newberriana, Spec. nov.
" 3, 4, 5. Goniopteris oblonga. Spec. nov.
" 3. Gives the normal form.
" 3^a. Enlarged pinnules of the same.
" 4. A portion of the upper part of the frond.
" 4^a. Enlarged pinnules from the lower part of Fig. 4.
" 4^b, 4^c. Enlarged portions of pinnæ from the upper part of Fig. 4.
" 5. Summit of a pinna enlarged.

PLATE XXXI.

Figs. 1, 2. Cymoglossa formosa, Spec. nov.
" 1^a. Enlarged pinnules of Fig. 1.
" 3. Cymoglossa breviloba, Spec. nov.
" 3^a. Enlarged pinnule of the same.
" 4. Cymoglossa lobata, Spec. nov.
" 4^a. Enlarged pinnule of the same.
" 5, 6. Cymoglossa obtusifolia, Spec. nov.
" 6. The basal portion of a pinna showing heteromorphous pinnules.
" 6^a. The same enlarged to show nervation.

PLATE XXXII.

Figs. 1–5. Alethopteris Virginiana, Spec. nov.
" 1. A compound pinna from the middle of the frond.
" 1^a. Enlarged pinnule of the same.
" 2. An ultimate pinna from the lower part of the frond showing undulate pinnules.

Fig.   3. The basal portion of an ultimate pinna, showing heteromorphous pinnules.

     4. Summit of a compound pinna.

     5. A fragment of a pinna, showing one of the variations of the pinnules.

Plate XXXIII.

Figs. 1-4. Alethopteris Virginiana.

     1. Pinnules showing what appears to be sori at the base of the pinnules.

     2. A fragment of the lower portion of a compound pinna, showing undulate basal pinnules.

     3. Summit of a pinna of the normal form.

     4. Fragment of a pinna with large pinnules of the normal form.

     4^a. Enlarged pinnule of the same.

     5, 6. Alethopteris gigas? Gein.

Plate XXXIV.

Figs. 1-8. Taeniopteris Newberriana, Spec. nov.

     1. Portion of a fructified frond.

     1^a. Fructification as seen on the upper surface of the frond.

     2. Basal portion of a fructified frond.

     3. Fragment showing imprints of the insertions of the sori.

     3^a. The imprints of the insertions of the sori enlarged.

     4, 5, 6. Portions of the sterile frond.

     5^a. A portion of the same enlarged.

     7. Basal portion of a frond which is possibly fructified higher up.

     8. A smaller sterile frond of probably the same species.

     9-9^a. Taeniopteris Lescuriana, Spec. nov. Fig. 9 shows a fragment of one side of the leaf.

PLATE XXXV.

Fig. 1. Rhacophyllum filiciforme, Var. majus.
" 2. Rhacophyllum laciniatum, Spec. nov.   Fig. 2.
Shows the plant on Pecopteris dentata.
" 3-4. Caulopteris eliptica, Spec. nov.
" 5. A portion showing the imprint of the outer
bark of the border around the scar, and a
part of the scar of Caulopteris gigantea.

PLATE XXXVI.

Caulopteris gigantea, Spec. nov.

PLATE XXXVII.

Fig. 1. Carpolithes marginatus, Spec. nov.
" 2. Guilielmites orbicularis, Spec. nov.
" 3. Sigillaria approximata, Spec. nov.
" 4. Undetermined plant. Apparently a portion
of a large flabellate leaf.
" 5. Impression of apparently the bark of an undetermined plant. Caulopteris?
" 6-7. Carpolithes bicarpus, Spec. nov.
" 8-9. Rhabdocarpus oblongus. Spec nov.
" 10. Cordaites crassinervis, Spec. nov.
" 11-12. Baiera Virginiana, Spec. nov.
" 12^a. A portion of 12, showing nervation.

PLATE XXXVIII.

Figs. 1-4. Saportea, Gen. nov.
" 1-3. Saportea Salisburioides, Spec. nov.
" 1^a. Shows the nervation of Fig. 1.
" 2. Gives a portion of a large leaf of the plant.
" 3. Shows in the right hand corner what seems
to be a portion of the terminal margin of
the leaf.
" 4. Saportea grandifolia, Spec. nov.
" 5, 5^a. A wing of a cockroach Gerablattina
balteata, Scudder.

INDEX TO PP.

Page.

ACROPTERIS cuneata (Schenk), ... 41
ADIANTUM, ... 100
   A. reniforme (L), ... 100
ALETHOPTERIS (Sternb.), ... 55,87,89,114,115
   A. (typical form), ... 14
   A. ambigua (Lx.), ... 17
   A. aquilina (Brongniart), ... 20,21
   A. gigas (Geinitz), ... 80,90,107,112,114—Pl. XXXIII, Figs. 5, 6.
   A. grandifolia (Newberry), ... 11,12, 14
   A. Helenae (Lesq.), ... 11, 12
   A. inflata (Lesq.), ... 89
   A. lonchitica (Brt. Var.), ... 11,12,13,14, 17
   A. nervosa (Brgt.), ... 17
   A. Pluckeneti (Schloth.), ... 17
   A. pteroides (Geinitz), ... 59
   A. Serlii (Brgt.), ... 17
   A. Sullivantii (Lx.), ... 17, 55
   A. Virginiana (spec. nov.), ... 88,96,107,114,—Pl. XXXII, Figs. 1-5.
                                              Pl. XXXIII, Figs. 1-4.
   A. (species nova, allied to *gigas* of Geinitz), ... 20
ANNULARIA (Sternb.), ... 38
   A. calamitoides (Schimp.), ... 112
   A. carinata (Guth.), ... 38,105,112
   A. longifolia (Brt.), ... 17,20,38,39,105
   A. minuta (Brt.), ... 39,106
   A. radiata (Brt.), ... 39,106
   A. sphenophylloides (Ung.), ... 17,20,39,106
APHLEBIA patens (Germ.), ... 97
ARCHÆOPTERIS (Daws.), ... 7,13,14
   A. Alleghanensis (*Cyclopteris* Allegh.) (Meek.), ... 6
   A. Boeckschiana (*Noeggerathia* Bœck.) (Gœpp.), ... 6
   A. Hibernica (*Palaeopteris* Hib.) (Forb.), ... 6
   A. Jacksoni (*Cyclopteris* Jacks.) (Daws.), ... 6
   A. obtusa (*Cyclopteris* of Dawson, *Noeggerathia* of Lesquereux)
      (Lx.), ... 6,7
ARTISIA transversa (Sternb.), ... 18

(135 PP.)

Page.

ASPLENITES Ottonis (Schenk), ... 59
ASPLENIUM, ... 72
ASTEROCARPUS (Weiss), ... 41
    A. Meriani (*Pecopteris*) (Heer,), ... 57
ASTEROPHYLLITES (Brt.), ... 35
    A. acicularis (Dawson), ... 11
    A. equisetiformis (Brt.), ... 17, 20
    A. foliosus (Lind. & Hutt.) 17
    A. longifolius, (Goepp.), ... 35
    A. rigidus (Brt.), ... 16
    A. sublævis (Lx.), ... 17
    A. species? (near equisetiformis), ... 20
BAIERA (Fr. Braun), ... 103,106
    B. digitata (Heer), ... 103,107,112
    B. Virginiana (sp. nov.), ... 103,107,112
    B. longifolia (Heer), ... 103
    B. Virginiana (sp. nov.)—Pl. XXXVII, Figs. 11, 12.
BLATTINA, species? (See *Gerablattina*), ... 104
BOCKSCHIA flabellata (Goep.), ... 34—Pl. XXXVIII, Fig. 8.
CALAMITES (Brt.), ... 34,35,114
    C. approximatus (Sternb.), ... 17
    C. cannæformis (Schloth.), ... 11, 20
    C. nodosus (Schloth.), ... 17
    C. ramosus (Artis.), ... 17
    C. Suckowii (Brt.), ... 17,35,105
CALAMOCLADUS (Schimp.), ... 35
CALAMOSTACHYS tuberculata, (Brt.), ... 17
CALLIPTERIDIUM (Weiss.), ... 55,56,58,60,61,62,71,88,115
    C. Dawsonianum (sp. nov.), . . 56,59,106—Pl. XIII, Figs. 1, 2. Pl. XIV,
                                           Fig. 1.
    C. grandifolium (sp. nov.), 58,106—Pl. XV, Figs. 1-4. Pl. XVI, Figs. 2-4.
    C. Mansfieldii (Lx.), ... 17
    C. odontopteroides (sp. nov.), ... 59,106—Pl. XVI, Fig. 1.
    C. oblongifolium (sp. nov.), ... 56,57,106—Pl. XII, Figs. 1-5.
    C. unitum (sp. nov.) . . . . 60, 106—Pl. XIV, Figs. 2, 3.
CALLIPTERIS (Brt.), ... 42,54,113,114
    C. conferta (Brt.), ... 42,54,105,112,113,116—Pl. XI, Figs 1-4.
CARDIOCARPUS (Brt.), ... 12
    C. mamillatus (Lx.), ... 18
CARDIOPTERIS frondosa (Schimp.), ... 7
CARPOLITHES (Sternb.), ... 98
    C. bicarpa (sp. nov.), ... 98,107—Pl. XXXVII, Figs. 6, 7.
    C. Canneltoni (Lx.), ... 18
    C. clypeiformis (Lx.), ... 18
    C. fasciculatus, (Lesq.), ... 98
    C. fraxiniformis, ... 18
    C. marginatus (sp. nov.), ... 98,107—Pl. XXXVIII, Fig. 1.
    C. multistriatus (Sternb), ... 18
    C. platimarginatus (Lx.), ... 18
    C. vesicularis (Lx.), ... 18

Page.

CAULOPTERIS (Lind. & Hut.), ... 95,107—Pl. XXXVII. Fig. 5.
 C. elliptica (sp. nov.), ... 95,107—Pl. XXXV, Figs. 3, 4.
 C. gigantea (sp. nov.), ... 95,107—Pl. XXXV, Fig. 5, Pl. XXXVI.
 C. macrodiscus (Brt.), ...95,107
 C. obtecta (Lx.), ... 17
 C. peltigera (Brgt.), ... 107
CLADOPHLEBIS (Schimp.), ... 61,71,115
CORDAISTROBUS (Grand 'Eury), ... 18
CORDAITES (Ung.), ... 12,18,97
 C. borassifolia (Ung.), ... 18,20
 C. communis (Lx.), ... 18
 C. costatus (Lx.), ... 18
 C. crassinervis (sp. nov.), ... 97, 107—Pl. XXXVII, Fig. 10.
 C. crassus (Lx.), ... 18
 C. diversifolius (Lx.), ... 18
 C. gracilis (Lx.), ... 18
 C. grandifolius (Lx.), ... 18
 C. Mansfieldi (Lx.), ... 18
 C. principalis (Daws.), ... 18
 C. reflexa (Lx.), ... 18
 C. Robii (Daws.), ... 12
 C. serpens (Lx.), ... 18
 C. validus (Lx.), ... 18
CORDIANTHUS fl. femina (2 species) (Antholithes), ... 18
 C. fl. masculina, ... 18
 C. gemmifer (Grd. Eury), ... 18
CARDIOPTERIS, ... 7
CYATHEITES Germari (Weiss), ... 68
 C. Pluckeneti (Brt.), ... 68, 69
 C. Schlotheimii, (Goep.), ... 62
 C. (Section Pecopteris), ... 61,113
CYCLOPTERIS (Brt.), ... 105
 C. Alleghanensis (Archæopteris all.) (Meek.), ... 6
 C. elegans (Brt.), ... 17
 C. fimbriata (Lx.), ... 17
 C. Jacksoni (Daws.), ... 6,7,12
 C. Lescuriana (Meek), ... 6,7
 C. obliqua (Lx.), ... 17
 C. trichomanoides (Lx.), ... 17
 C. undans (Lx.), ... 17
 C. valida (Daws.), ... 7
 C. Virginiana (Meek), ... 6,7
 (also see Archæopteris obtusa.)
CYMOGLOSSA (Schimp.), ... 81,84,113
 C. breviloba (sp. nov.), ... 86,107—Pl. XXXI, Fig. 3.
 C. formosa (sp. nov.), ... 86,107—Pl. XXXI, Figs. 1, 2.
 C. Gœppertiana (Schimp.), ... 86
 C. lobata (sp. nov.), ... 87,107—Pl. XXXI, Fig. 4.
 C. obtusifolia (sp. nov.), ... 85,107—Pl. XXXI, Figs. 5, 6.
CYPRIS, ... 116

Page

DESMIOPHYLLUM gracile (Lx.), ... 18
DICRANOPHYLLUM species, ... 18
    D. dimorphum (Lx.), ... 18
DICTYOPTERIS (Gutb.), ... 50
    D. neuropteroides (Von Röhl), ... 50
    D. obliqua (Bunb.), ... 17
EQUISETIDES (Schimp.), ... 33,34
    E. elongatus (sp. nov.), ... 33,105,115,121—Plate I, Figs. 1-4.
    E. rugosus (Schimp.), ... 33,105,121—Pl. I, Fig. 6.
    E. striatus (sp. nov.), ... 34,105,121—Plate I, Fig. 5.
EQUISETITES infundibuliformis (Schimp.), ... 17
EQUISETUM triphyllum (Heer), ... 115
EREMOPTERIS (Schimp.), ... 94
ESTHERIA, ... 116
GENUS ?, ... 97
GERABLATTINA balteata (Scudder), ... 104—Pl. XXXVIII, Fig. 5.
GINGKO (see Salisburia), ... 116
GLEICHENITES Noesii, (Goep.), ... 94
GONIOPTERIS (Presl.), ... 80,81,84,85,113
    G. arguta (Schimp.), ... 82,83,84,87,107,111
    G. elegans (Schimp.), ... 82,107
    G. elliptica (sp. nov.), ... 83,107—Pl. XXXI, Fig. 1a.
    G. emarginata (Schimp.), ... 80,82,86,107
    G. longifolia, (Schimp.), ... 82,107
    G. Newberriana (sp. nov.), ... 84,107—Pl. XXX.
    G. oblonga (sp. nov.), ... 83,107—Pl. XXX, Figs. 3-5.
    G. species ?, ... 83
GUILIELMITES (Geinitz), ... 99
    G. orbicularis (sp. nov.), ... 99,107,114—Pl. XXXVII, Fig. 2.
    G. permianus (Goepp.), ... 107,114
HYMENOPHYLLITES expansus, ... 17
    H. Gutbierianus (Presl.), ... 17
    H. laceratus (Lx.), ... 17
    H. lactuca (Gutb.), ... 17
JEANPAULIA, ... 103
    J. longifolia (Heer), ... 103
KNORRIA acicularis (Goep.), ... 8
LEPIDODENDRON (Sternb.), ... 7,12,114
    L. modulatum (Lx.), ... 17
    L. obovatum (Sternb.), ... 17
    L. quadratum (Lx.), ... 17
    L. selaginoides (Sternb.), ... 11
    L. Sternbergii (Brt.), ... 6,16,17
    L. Veltheimianum (Sternb.), ... 6
    L. species ?, ... 6,8
LEPIDOPHLOIOS LARICINUS (Sternb.), ... 18
LEPIDOPHYLLUM auriculatum (Lx.), ... 17
    L. foliaceum (Lx.), ... 17
    L. Mansfieldi (Lx.), ... 17
    L. undulatum (Lx.), ... 17
    L. species ? 16

Page.

LEPIDOSTROBUS ornatus (Brt.), .16,17
  L. variabilis (L. & H.), ... 17
LEPIDOXALON anomalum (Lx.), ... 18
LESCUROPTERIS Moorii (Schimp.), ...20,21
MACROTAENIOPTERIS (Schimp.), ... 93
  M. gigantea (Schenk), ... 91
  M. Rogersi, (Schimp), ... 92
MEGALOPTERIS (Daws.), ...13,14
  M. Hartii, (Andr.), ... 11
  M. Sewellensis (Font.), ...11,12
MERIANIOPTERIS (Heer), ...78,115
  M. angusta (Heer), ... 56
MIXONEURA (sub-genus; Weiss), ...53,114,115
NEMATOPHYLLUM (gen. nov.), ...35,113
  N. angustum, ... 35,105,121—Pl. II, Figs. 1–5.
NEUROPTERIDIUM (sub-genus; Schimper), ... 51
NEUROPTERIS (Brt.), ... 7,14,46,49,50,55,58,61,115
  N. acutifolia (Brt.), ...16,20
  N. angustifolia (Brt.), ... 17
  N. auriculata (Brt.), ...50,106
  N. Clarksoni (Lesq.), ...16,17
  N. cordata (Brt.), ... 48,51,106
  N. cordato-ovata (Weis), ... 61
  N. cordifolia (Lx.), ... 17
  N. crenulata (Brt.), ... 17
  N. Dawsoni (Hartt.), ... 12
  N. dictyopteroides (sp. nov.), ... 14,72,106—Pl. VIII, Figs. 3–5.
  N. Dufresnoyi (Brt.), ... 115
  N. frimbriatus (Lesq.), ...51,106,111
  N. flexuosa(Brt.), ... 7,12,16,17,20,47,48,49,51,106—Pl. VIII, Fig. 6.
  N. flexuosa, Var. longifolia, ...49,106
  N. Grangeri (Brt.), ... 20
  N. heterophylla (Brt.), ... 16
  N. hirsuta (Lesq.), . . . 16,17,20,47,48,106,111,112,116—Pl. VIII,Fig.8.
  N. Loschii (Brt.), ...17,20
  N. odontopteroides (sp. nov.), ... 50,106—Pl. IX, Fig. 16.
  N. platynervis (sp. nov.), ... Pl. VIII, Fig. 2.
  N. plicata (Sternb.), ... 17
  N. rarinervis (Bunb.), ...16,20
  N. Rogersi (Lesq.), ... 52
  N. Smithiana (Lesq.), ...11,12,14
  N. tenuifolia (Brt.), ...11,17
  N. vermicularis (Lx.), ... 17
  N. Villersii (Brt.), ... 50
  N. species? . . . . 51—Pl. X, Fig. 11.
NOEGGERATHIA (See Archæopteris obtusa.)
  N. Boekschiana, (Archæopteris B.), ... 6
  N. dispar (Dawson), ... 100
  N. obtusa (Lesq.), ... 7
ODONTOPTERIS (Brt.), ...52,60,114,115

Page.

O. alpina (Heer), ... 53
O. densifolia (sp. nov.), ... 54,106
O. Dufresnoyi (Brt.), ... 51
O. gracillima (Newb.), ... 11
O. nervosa (sp. nov.), ... Pl. X, Figs. 1-2, 52,106
O. neuropteroides (Newb.), ...11,14
O. obtusa (Naum.), ... 52
O. nervosa (sp. nov.), ... 106,112,114—Pl. X, Figs. 1-2.
O. obtusiloba, Var. varinervis, ... 52—Pl. X, Fig. 4.
O. pachyderma (sp. nov.), ... 53,106—Pl. X, Figs. 10.
O. Schlotheimii (Brt.), ... 17
O. subcuneata (Bunb.), ... 16
O. (sp. nov. allied to obtusiloba), ... 20
OLEANDRA nereiformis (Presl.), ... 93
OLEANDRIDIUM (Schimp.), ...90,93
ORTHOGONIOPTERIS (Andr.), ... 12
PACHYPTERIS (Brt.), ...76,115
PALAEOPTERIS Hibernica (Archæopteris H.) (Forb.), ... 6
PECOPTERIDIUM (suggested genus,) 61
PECOPTERIS (Brt.), ...7,14,20,40,43,46,47,55,56,61,62,63,71,114,115
P. (section Cyatheides), ... 61,115
P. acrostichoides (Schimper), ... 77
P. adiantoides (L. & H.), ... 72
P. angustipinna (sp. nov.), ... 76,106—Pl. XXVII, Figs. 1-3.
P. arborescens (Schloth.), ... 16,62,63,77,78,106,115
P. arborescens, Var. integripinna 63,106
P. arguta (Brt.), .83,84
P. asplenioides (sp. nov.), ... 72,106—Pl. XXV, Fig. 1.
P. Bredovi (Germ.), ... 71
P. Bucklandi (Brt.), ... 20
P. Candolleana (Brt.), ... 20,21,63,106—Pl. XX, Figs. 1-3.
P. choerophylloides (Brt.), ... 17
P. concinna (Lesq.), ... 73
P. constricta, ... 68
P. cristata, (Brt.), ...45,46
P. dentata, (Brt.), 67,76,94,106—Pl. XXII, Figs. 1-5. Pl. XXXV, Fig. 2.
P. dentata. Var. crenata, ... 66
P. dentata. Var. parva, ... 67
P. dentata. Var. plumosa, ... 20,66—Pl. XXII, Fig. 2.
P. elliptica (Bunb.), ... 64,82,106,111—Pl. XVII, Fig. 1.
P. Germari (Weiss), Font. & White), .68,69,70,106—Pl. XIX, Fig. 1-7.
P. Germari Var. crassinervis, ... 70,106—Pl. XX, Fig. 5.
P. Germari Var. cuspidata, ... 76,106—Pl. XX, Fig. 4.
P. Goepperti (Morris), ... 84
P. goniopteroides (sp. nov.), ... 80,106—Pl. XXV, Fig. 2.
P. Heeriana (sp. nov.), ... 77,106—Pl. XXV, Figs. 3-7.
P. hemiteloides (Brt.), ... 17
P. imbricata (sp. nov.), ... 72,106—Pl. XXIII, Fig. 1.
P. inclinata (sp. nov.), ... 80,106—Pl. XXIX, Fig. 4.
P. lanceolata (sp. nov.), ... 79,106—Pl. XXIX, Figs. 7-9.

Page.

P. latifolia (sp. nov.), ... 79,106—Pl. XXIX, Figs. 5,6.
P. Merianiopteroides (sp. nov.), . . . 78,106,115—Pl. XXIX, Figs. 1,2.
P. Meriani (Heer), ... 57
P. microphylla (Lx.), ... 17
P. Miltoni (Artis), ... 65,106—Pl. XXIII, Figs. 2,3.
P. Miltoni Var. polymorpha, ... 66
P. muricata (Brt.), ... 11,13,14
P. nervosa (Brt.), ... 11,12,13,14
P. notata (Lesq.), ... 20,68,106,111
P. sub-falcata (sp. nov.), ... 70,80,106
P. oreopteridia (Brt.), ... 64,74,106
P. ovoides (sp. nov.), ... 79,106—Pl. XXIX, Fig. 3.
P. pachypteroides (sp. nov.), ... 76,106,115—Pl. XXVI, Figs. 1-4.
P. pennaeformis (Brt.), ... 65,106
P. pennaeformis Var. latifolia, ... 65,106—Pl. XVII, Figs. 4,5.
P. pinnatifida (Gutb.), (Gein.), ... 70
P. platynervis (sp. nov.), ... 73,106—Pl. XVIII, Figs. 1-6.
P. Pluckeneti (Brt.), ... 20,67,68,69,106—Pl. XXI, Figs. 4,5.
P. Pluckeneti Var. constricta, ... 68,106—Pl. XXI, Fig. 3.
P. plumosa (Brgt.), ... 17,67
P. polymorpha (Brt.), ... 17,66
P. pteroides (Brt.), ... 20,59,67,71,106
P. rarinervis (sp. nov.), ... 71,106—Pl. XX, Figs. 6-8.
P. rotundifolia (sp. nov.), ... 73,106—Pl. XXIV, Fig. 6.
P. rotundiloba (sp. nov.), ... 74,106—Pl. XVII, Fig. 2.
P. Schimperiana (sp. nov.), ... 75,106—Pl. XXIV, Figs. 1-5.
P. Sillimani (Brgt.), ... 17
P. spinulosa (Lesq.), ... 20
P. squamosa (Lx.), ... 17
P. subfalcata (sp. nov.), ... Pl. XXI, Figs. 1,2.
P. Sulziana (Brt.), ... 75
P. tenuinervis (sp. nov.), ... 77,94,106—Pl. XXVIII, Figs. 1-4.
P. triassica (Heer), ... 53
P. truncata (Lx.), ... 17
P. unita (Brt.), ... 79
P. villosa (Brt.), ... 16
P. Whitbiensis (Brt.), ...
P. Williamsoni (Brt.), ... 77
P. species? . . . . 80,81,82—Pl. XVII, Figs. 3,6; Pl. XXIV, Fig. 7; Pl.
    XXVII, Figs. 4,5,6.
P. (See Asterocarpus Merianus.)
PHYLLOTHECA (Brt.), ... 34
PINNULARIA capillacea (Ll. & Hutt.), ... 18
PSYGMOPHYLLUM, ... 97
PTEROPHYLLUM, ... 93
RHABDOCARPUS (Goep. and Berg.), ... 12,98
    R. amygdalaeformis (Goepp.), ... 18
    R. Boehsianus, ... 18
    R. clavatus (Sternb.), ... 18
    R. oblongatus (sp. nov.), ... 98,107—Pl. XXXVII, Figs. 8,9.

Page.

RHACOPHYLLUM (Schimp.), ... 93
  R. filiciforme (Schimp.), ... 20,93,94,107
  R. filiciforme Var. *majus*, ... 93—Pl. XXXV, Fig. 1.
  R. laciniatum (sp. nov.), ... 94,107—Pl. XXXV, Fig. 2.
  R. lactuca (Sternb.), ... 94,107
  R. speciocissimum (Schimp.), ... 94,107
RHIZOMORPHA sigillariae (Lx.), ... 18
SALISBURIA (Gingko), ... 100,101,102,103,116
SAPORTÆA (gen. nov.), ... 99,115—Pl. XXXVIII, Figs. 1-4.
  S. grandifolia (sp. nov.), ... 101,102,107—Pl. XXXVIII, Fig. 4.
  S. Salisburioides (sp. nov.), . . . 102,103,107—Pl. XXXVIII, Figs. 1-3.
SCHIZONEURA, ... 35,113,114
  S. Meriani (Heer), ... 35,113
SCHIZOPTERIS lactuca (Presl.), ... 94,107
SCOLOPENDRIUM, ... 48
  S. vulgare (Lx.), ... 48
SIGILLARIA (Brt.), ... 7,12,96,114
  S. alternans (Ll. & Hutt.), ... 18
  S. approximata (sp. nov.), ... 96,107—Pl. XXXVII, Fig. 3.
  S. Brardii (Brt.), ... 97,107,114
  S. elliptica (Brt.), ... 18
  S. mamillaris (Brt.), ... 18
  S. Menardi (Brt.), ... 96
  S. monostigma (Lx.), ... 18
  S. pes-capreoli (Gein.), ... 20
  S. reniformis (Brt.), ... 18
  S. sculpta (Lx.), ... 18
  S. tessellata (Brt.), ... 18
SOLENOMYA, ... 117
  S. permiensis (White), ... 117
SPHENOPHYLLUM (Brt.), ... 36,37
  S. angustifolium (Germ.), ... 37
  S. densifoliatum (sp. nov.), ... 37,38, 105,121—Pl. I, Fig. 8.
  S. emarginatum, ... 17
  S. filiculmis (Lesq.), ... 20,37,38,105,111,121—Plate I, Fig. 8.
  S. latifolium, (sp. nov.), ... 36,105,121—Pl. I, Figs. 10, 11.
  S. longifolium (Germ.), ... 17,36,38,105
  S. oblongifolium (Germ.), ... 38,105
  S. Schlotheimii (Brt.), ... 16,17
  S. tenuifolium (sp. nov.), ... 38,105,121—Pl. I, Fig. 9.
  S. trifoliatum (Lesq.), ... 20
SPHENOPTERIS (Brt.) 7,14,40,43,46,47,113,115
  S. (*Pecopteroid* Section), ... 44,45
  S. acrocarpa (sp. nov.), 40,106,115,122—Pl. III, Figs. 1-3. Pl. IV, Figs. 1-5.
  S. adiantoides (L. & H.), ... 11
  S. Artemisiaefolia (Sternb.), ... 17
  S. auriculata (sp. nov.), ... 42,106—Pl. VII, Figs. 3, 4.
  S. coriacea (sp. nov.), ... 41,54,106,112,113,116,123—Pl. V, Fig. 5.
  S. cristata (Brt.), ... 43
  S. dentata (sp. nov.), ... 42,106,123—Pl. V, Figs. 7, 8.

Page.

S. denticulata (Brt.), ... 41
S. flaccida (Lx.), ... 8
S. foliosa (sp. nov.), ... 44,106,123—Pl. V, Figs. 9-11.
S. furcata (Brt.), ... 20
S. hastata (sp. nov.) . . . . 46,106—Pl. VII, Fig. 7.
S. Hoeninghausi (Brt.), ... 11,12
S. integra (Goep.), ... 65
S. latifolia (Brt.), ... 14,40
S. Lescuriana (sp. nov.), . . 44,106—Pl. VI, Fig. 1. Pl. VII, Figs. 1, 2.
S. Lesquereuxii (Newb.), ... 42
S. lyratifolia (Weiss), ... 45,105,113
S. macilenta (L. & H.), ... 7,11,12,14,40
S. minuti-secta (sp. nov.), ... 20,43,106,115—Pl. V, Fig. 1-4.
S. mixta (Schimp.), ... 17
S. Newberryi (Lx.), ... 17
S. Nummularia (Gutb.), ... 70
S. obtusiloba (Brt.), ... 11,13,14
S. oxydata (Goep.), ... 42,106,113
S. pachynervis (sp. nov.), ... 46,106—Pl. VII, Figs. 5, 6.
S. species ?, ... 42
S. Sarana (Weiss), ... 42
SPHENOPTERIS Pecopterides (suggested sub-genus), ... 46
SPIROPTERIS villosa (Lx.), ... 17
STEMMATOPTERIS Mansfieldi (Lx.), ... 17
STIGMARIA minuta (Goep.), ... 8
S. ficoides (Brt.), ... 18
S. Wolkmannia, ... 8
SYRINGODENDRON cyclostigma (Brt.), ... 18
S. pes-capreoli (Brt.), ... 18
TAENIOPHYLLUM contextum (Lx.), ... 18
T. decurrens (Lx.), ... 18
T. deflexum (Lx.), ... 18
TAENIOPTERIS (Brt.), ... 12,90,93,101
T. coriacea, (Goep.), ... 93,112
T. Lescuriana (sp. nov.), ... 91,93,107,112—Pl. XXXIV, Fig. 9.
T. multinervis (Weiss), ... 91,107,112
T. Newberriana (sp. nov.), ... 91,93,107,112—Pl. XXXIV, Figs. 1-8.
T. Newberriana Var anguste, ... 93
T. Smithii (Lesq.), ... 12
T. vittata (Brt.), ... 91,93,107
THYRSOPTERIS (Heer), ... 44,115
TRIGONOCARPUS (Brt.), ... 12
T. Daviesii, ... 18
TRIPHYLLOPTERIS, ... 7,14
T. Lescuriana (Meek.), ... 6
T. Virginiana (Meek.), ... 6
T. (species undescribed), ... 6
ULODENDRON majus L. & H.), ... 8
WHITTLESEYA elegans (Newb.), ... 13
ZONARITES digitatus (Gein.), ... 103

Second Geological Survey of Pennsylvania.

Reports for 1874, 1875, 1876, 1877, 1878, and 1879.

The following Reports are issued for the State by the Board of Commissioners, at Harrisburg, and the prices have been fixed as follows, in accordance with the terms of the act:

PRICES OF REPORTS.

A. Historical Sketch of Geological Explorations in Pennsylvania and other States. By J. P. Lesley. With appendix, containing Annual Reports for 1874 and 1875; pp. 256, 8vo. Price in paper, $0 25; postage, $0 06. Price in cloth, $0 50; postage, $0 10.

B. Preliminary Report of the Mineralogy of Pennsylvania—1874. By Dr. F. A. Genth. With appendix on the hydro-carbon compounds, by Samuel P. Sadtler. 8vo., pp. 206, with *map* of the State for reference to counties. Price in paper, $0 50; postage, $0 08. Price in cloth, $0 75; postage, $0 10.

B.² Preliminary Report of the Mineralogy of Pennsylvania for 1875. By Dr. F. A. Genth. Price in paper, $0 65; postage, $0 02.

C. Report of Progress on York and Adams Counties—1874. By Persifor Frazer, Jr. 8vo., pp. 198, illustrated by 8 *maps* and *sections* and other illustrations. Price in paper, $0 85; postage, $0 10. Price in cloth, $1 10; postage, $0 12.

CC. Report of Progress in the Counties of York, Adams, Cumberland, and Franklin—1875. Illustrated by *maps* and *cross-sections*, showing the Magnetic and Micaceous Ore Belt near the western edge of the Mesozoic Sandstone and the two Azoic systems constituting the mass of the South Mountains, with a preliminary discussion on the Dillsburg Ore Bed and catalogue of specimens collected in 1875. By Persifor Frazer, Jr. Price, $1 25; postage, $0 12.

D. Report of Progress in the Brown Hematite Ore Ranges of Lehigh County—1874, with descriptions of mines lying between Emaus, Alburtis, and Foglesville. By Frederick Prime, Jr. 8vo., pp. 73, with a contour-line *map* and 8 *cuts*. Price in paper, $0 50; postage, $0 04. Price in cloth, $0 75; postage, $0 06.

DD. The Brown Hematite Deposits of the Siluro-Cambrian Limestones of Lehigh County, lying between Shimersville, Millerstown, Schencksville, Balletsville, and the Lehigh river—1875-6. By Frederick Prime, Jr. 8 vo., pp. 99, with 5 *map-sheets* and 5 *plates*. Price, $1 60; postage, $0 12.

E. Special Report on the Trap Dykes and Azoic Rocks of Southeastern Pennsylvania, 1875; Part I, Historical Introduction. By T. Sterry Hunt. 8 vo., pp. 253. Price, $0 48; postage, $0 12.

F. Report of Progress in the Juniata District on Fossil Iron Ore Beds of Middle Pennsylvania. By John H. Dewees. With a report of the Aughwick Valley and East Broad Top District. By C. A. Ashburner. 1874-8. Illustrated with 7 *Geological maps* and 19 *sections*. 8 vo., pp. 305. Price, $2 55; postage, $0 20.

G. Report of Progress in Bradford and Tioga Counties—1874-8. I. Limits of the Catskill and Chemung Formation. By Andrew Sherwood. II. Description of the Barclay, Blossburg, Fall Brook, Arnot, Antrim, and Gaines Coal Fields, and at the Forks of Pine Creek in Potter County. By Franklin Platt. III. On the Coking of Bituminous Coal. By John Fulton. Illustrated with 2 colored *Geological* county *maps*, 3 *plates* and 35 *cuts*. 8 vo., pp. 271. Price, $1 00; postage $0 12.

H. Report of Progress in the Clearfield and Jefferson District of the Bituminous Coal Fields of Western Pennsylvania—1874. By Franklin Platt. 8vo., pp. 296, illustrated by 139 *cuts*, 8 *maps*, and 2 *sections*. Price in paper, $1 50; postage, $0 15. Price in cloth, $1 75; postage, $0 15.

HH. Report of Progress in the Cambria and Somerset District of the Bituminous Coal Fields of Western Pennsylvania—1875. By F. and W. G. Platt. Pp. 194, illustrated with 84 *wood-cuts* and 4 *maps* and *sections*. Part I. Cambria. Price, $1 00; postage, $0 12.

HHH. Report of Progress in the Cambria and Somerset District of the Bituminous Coal Fields of Western Pennsylvania—1876. By F. and W. G. Platt. Pp. 348, illustrated by 110 *woodcuts* and 6 *maps* and *sections*. Part II. Somerset. Price, $0 85; postage, $0 18.

HHHH. Report of Progress in Indiana County—1877. By W. G. Platt. Pp. 316. With a colored map of the county. Price, $0 80; postage, $0 14.

I. Report of Progress in the Venango County District—1874. By John F. Carll. With observations on the Geology around Warren, by F. A. Randall; and Notes on the Comparative Geology of North-eastern Ohio and Northwestern Pennsylvania, and Western New York, by J. P. Lesley. 8 vo., pp. 127, with 2 *maps*, a long *section*, and 7 *cuts* in the text. Price in paper, $0 60; postage, $0 05. Price in cloth, $0 85; postage, $0 08.

II. Report of Progress, Oil Well Records, and Levels—1876-7. By John F. Carll. Pp. 398. Published in advance of Report of Progress, III. Price, $0 60; postage, $0 18.

J. Special Report on the Petroleum of Pennsylvania—1874, its Production, Transportation, Manufacture, and Statistics. By Henry E. Wrigley. To which are added a Map and Profile of a line of levels through Butler, Armstrong, and Clarion Counties, by D. Jones Lucas; and also a Map and Profile of a line of levels along Slippery Rock Creek, by J. P. Lesley. 8 vo., pp. 122; 5 *maps* and *sections*, a *plate* and 5 *cuts*. Price in paper, $0 75; postage, $0 06. Price in cloth, $1 00; postage, $0 08.

K. Report on Greene and Washington Counties—1875, Bituminous Coal Fields. By J. J. Stevenson, 8 vo., pp. 420, illustrated by 3 *sections* and 2 county *maps*, showing the depth of the Pittsburg and Waynesburg coal bed, beneath the surface at numerous points. Price in paper, $0 65; postage, $0 16. Price in cloth, $0 90; postage, $0 18.

KK. REPORT OF PROGRESS IN THE FAYETTE AND WESTMORELAND DISTRICT OF THE BITUMINOUS COAL FIELDS OF WESTERN PENNSYLVANIA—1876. By J. J. Stevenson; pp. 437, illustrated by 50 wood-cuts and 3 county maps, colored. Part I. Eastern Allegheny County, and Fayette and Westmoreland Counties, west from Chestnut Ridge. Price, $1 40; postage, $0 20.

KKK. REPORT OF PROGRESS IN THE FAYETTE AND WESTMORELAND DISTRICT OF THE BITUMINOUS COAL FIELDS of Western Pennsylvania—1877. By J. J. Stevenson. Pp. 331. Part II. The LIGONIER VALLEY. Illustrated with 107 wood-cuts, 2 plates, and 2 county maps, colored. Price, $1 40; postage, $0 16.

L. 1875—SPECIAL REPORT ON THE COKE MANUFACTURE OF THE YOUGHIOGHENY RIVER VALLEY IN FAYETTE AND WESTMORELAND COUNTIES, with Geological Notes of the Coal and Iron Ore Beds, from Surveys, by Charles A. Young; by Franklin Platt. To which are appended: I. A Report on Methods of Coking, by John Fulton. II. A Report on the use of Natural Gas in the Iron Manufacture, by John B. Pearse, Franklin Platt, and Professor Sadtler. Pp. 252. Price, $1 00; postage, $0 12.

M. REPORT OF PROGRESS IN THE LABORATORY OF THE SURVEY AT HARRISBURG—1874-5, by Andrew S. McCreath. 8 vo., pp. 105. Price in paper, $0 50; postage, $0 05. Price in cloth, $0 75; postage, $0 08.

MM. SECOND REPORT OF PROGRESS IN THE LABORATORY OF THE SURVEY at Harrisburg, by Andrew S. McCreath—1876-8, including I. Classification of Coals, by Persifor Frazer, Jr. II. Firebrick Tests, by Franklin Platt. III. Notes on Dolomitic Limestones, by J. P. Lesley. IV. Utilization of Anthracite Slack, by Franklin Platt. V. Determination of Carbon in Iron or Steel, by A. S. McCreath. With 3 indexes, plate, and 4 page plates. Pp. 438. Price in cloth, $0 65; postage, $0 18.

N. REPORT OF PROGRESS—1875-6-7. Two hundred Tables of Elevation above tide level of the Railroad Stations, Summits and Tunnels; Canal Locks and Dams, River Riffles, &c., in and around Pennsylvania; with map; pp. 279. By Charles Allen. Price, $0 70; postage, $0 15.

O. CATALOGUE OF THE GEOLOGICAL MUSEUM—1874-5-6-7. By Charles E. Hall. Part I. Collection of Rock Specimens. Nos. 1 to 4,264. Pp. 217. Price, $0 40; postage, $0 10.

P. 1879—ATLAS OF THE COAL FLORA OF PENNSYLVANIA AND OF THE CARBONIFEROUS FORMATION THROUGHOUT THE UNITED STATES. 87 plates with explanations. By Leo Lesquereux. Price, $3 35; postage, $0 22.

Q. REPORT OF PROGRESS IN THE BEAVER RIVER DISTRICT OF THE BITUMINOUS COAL FIELDS OF WESTERN PENNSYLVANIA. By I. C. White; pp. 337, illustrated with 3 Geological maps of parts of Beaver, Butler, and Allegheny Counties, and 21 plates of vertical sections—1875. Price, $1 40; postage, $0 20.

QQ. REPORT OF PROGRESS IN 1877. The Geology of Lawrence County, to which is appended a Special Report on the CORRELATION OF THE COAL MEASURES in Western Pennsylvania and Eastern Ohio. 8 vo., pp. 336, with a colored Geological Map of the county, and 134 vertical sections. By I. C. White. Price, $0 70; postage, $0 15.

V. REPORT OF PROGRESS—1878. Part I. The Northern Townships of Butler county. Part II. A special survey made in 1875, along the Beaver and Shenango rivers, in Beaver, Lawrence and Mercer Counties. 8 vo., pp. 248, with 4 maps, 1 profile section and 154 vertical sections. By H. Martyn Chance. Price, $0 70; postage, $0 15.

Other Reports of the Survey are in the hands of the printer, and will soon be published.

The sale of copies is conducted according to Section 10 of the Act, which reads as follows:

* * * "Copies of the Reports, with all maps and supplements, shall be donated to all public libraries, universities, and colleges in the State, *and shall be furnished at cost of publication to all other applicants for them.*"

Mr. F. W. FORMAN is authorized to conduct the sale of reports; and letters and orders concerning sales should be addressed to him, at 223 Market street, Harrisburg. Address general communications to WM. A. INGHAM, Secretary.

By order of the Board,

WM. A. INGHAM,
Secretary of Board.

Rooms of Commission and Museum: Address of Secretary:
223 Market Street, Harrisburg. *223 Market Street, Harrisburg.*

(4)

1a

1

5

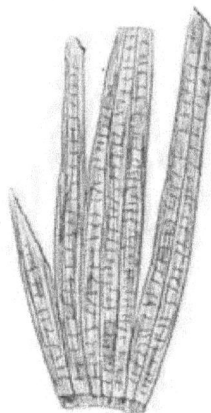

6

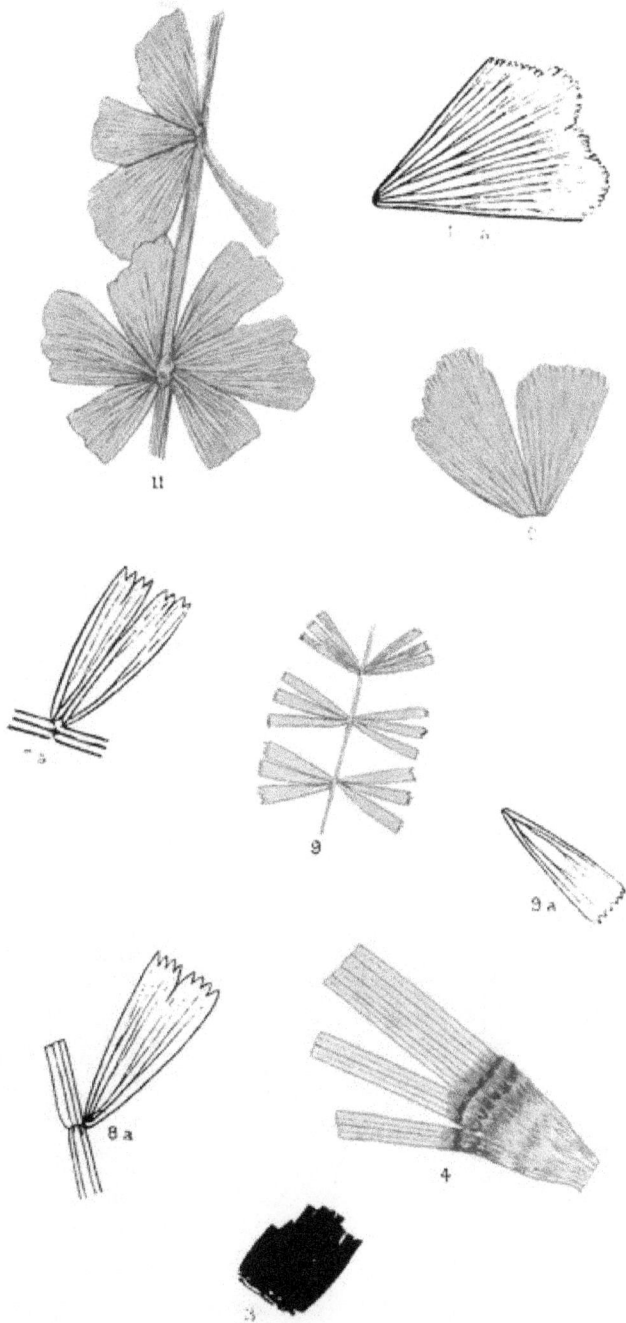

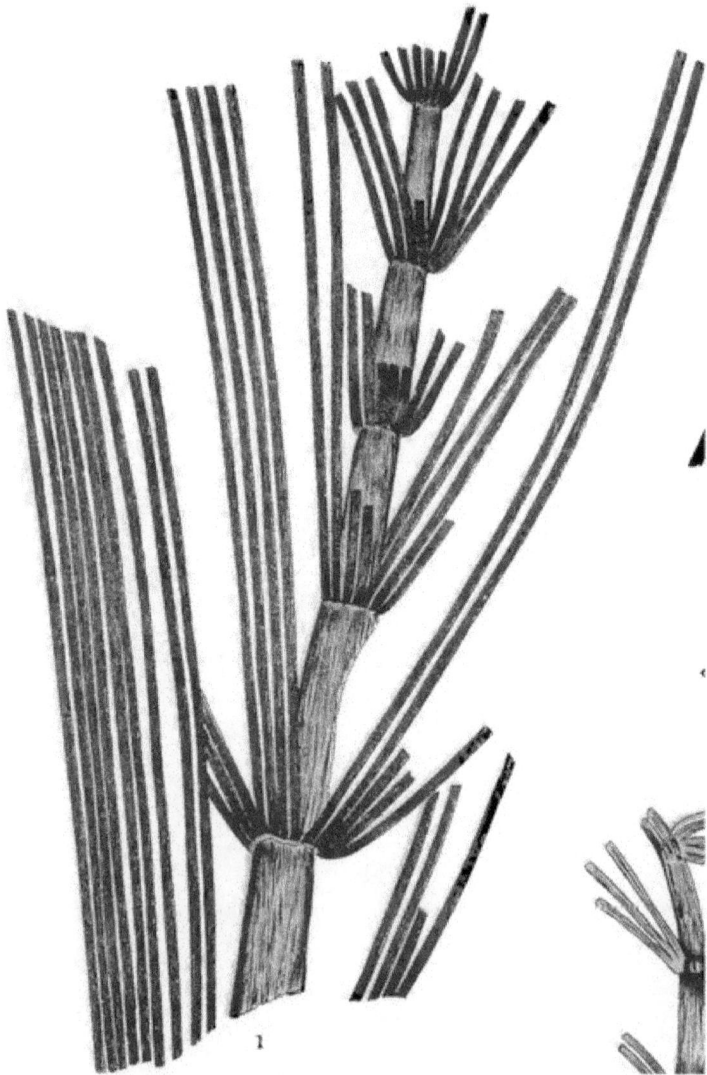

1

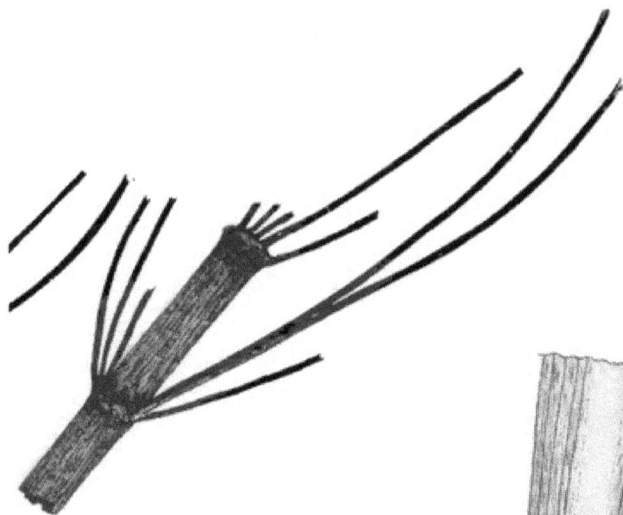

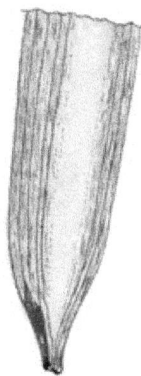

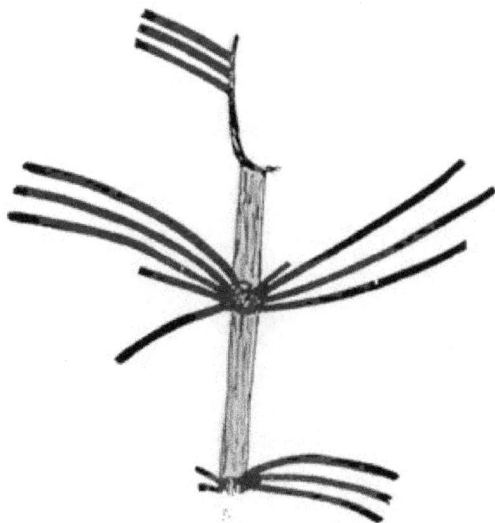

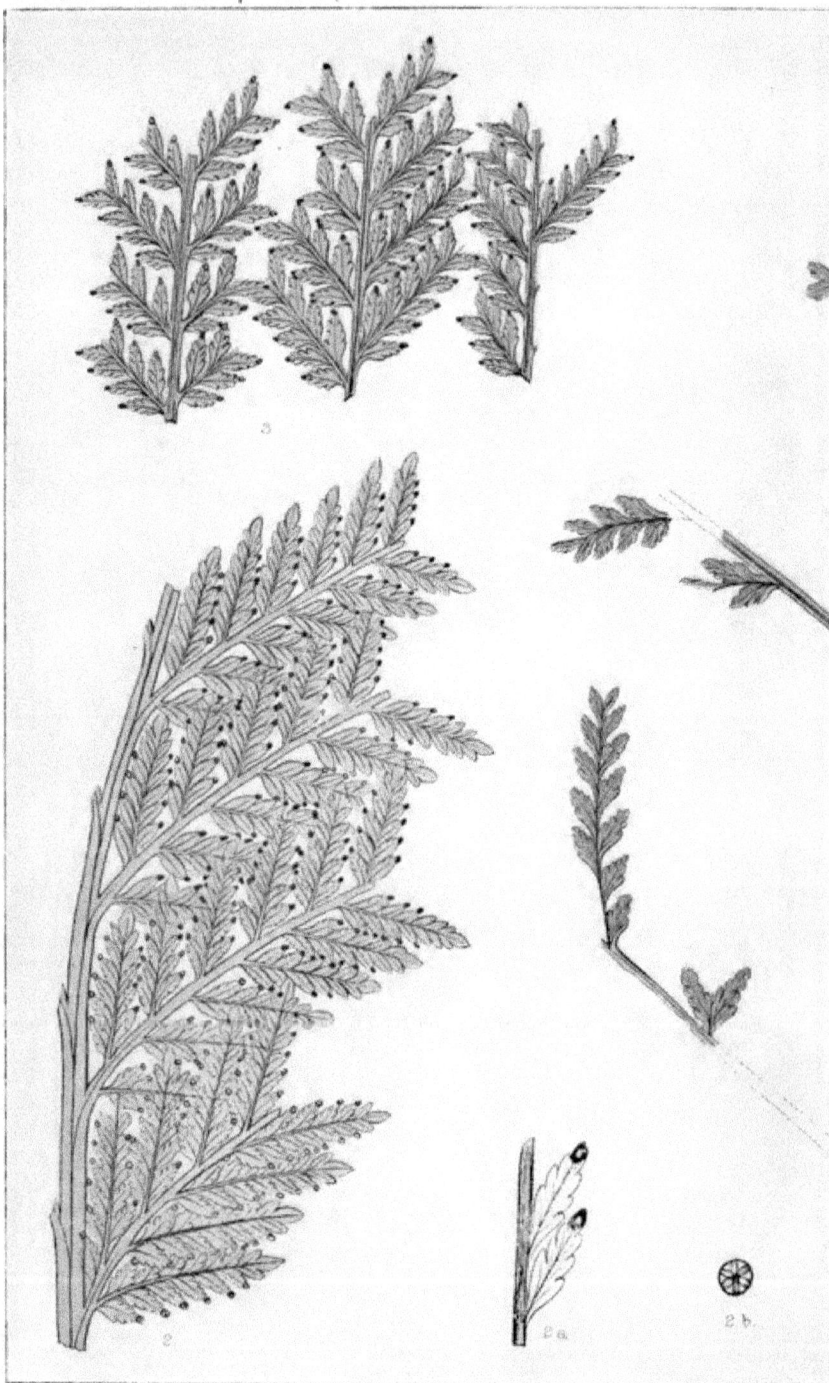

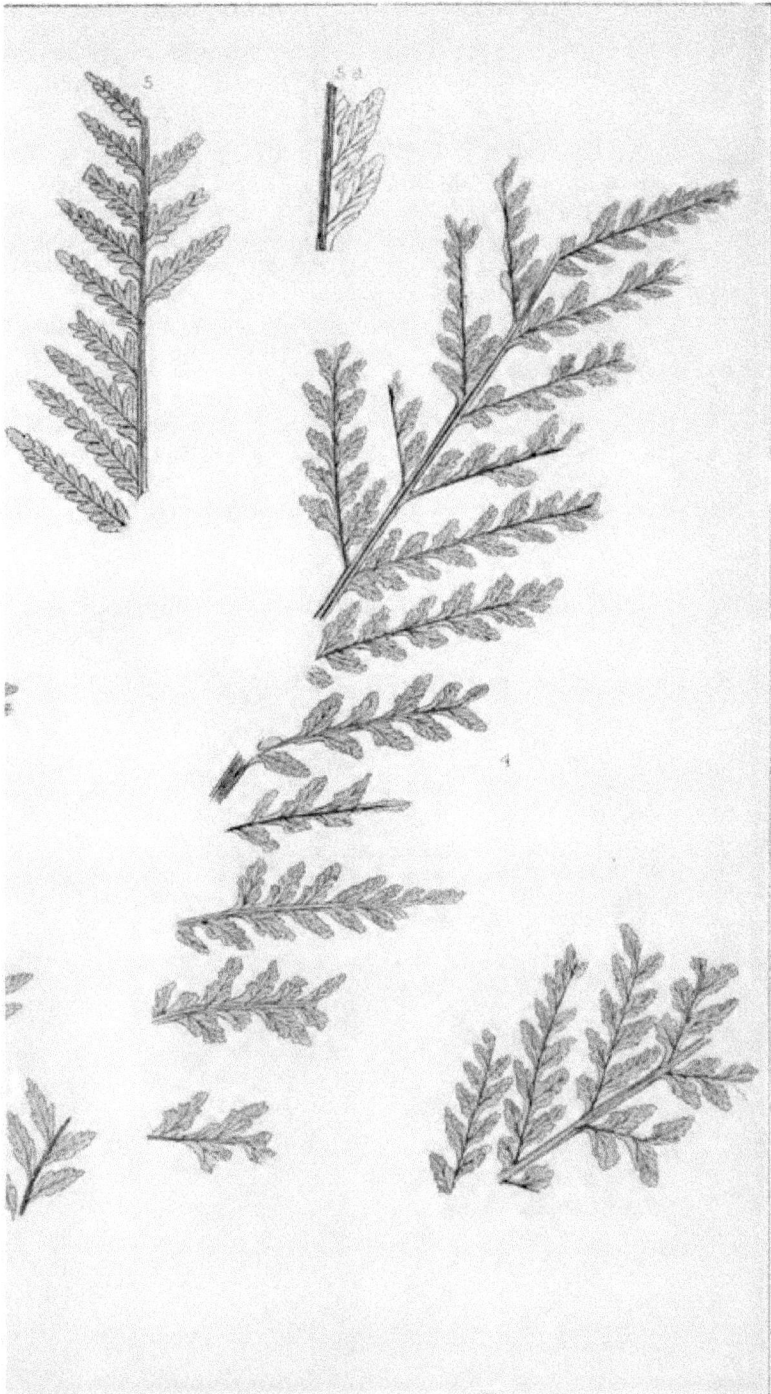

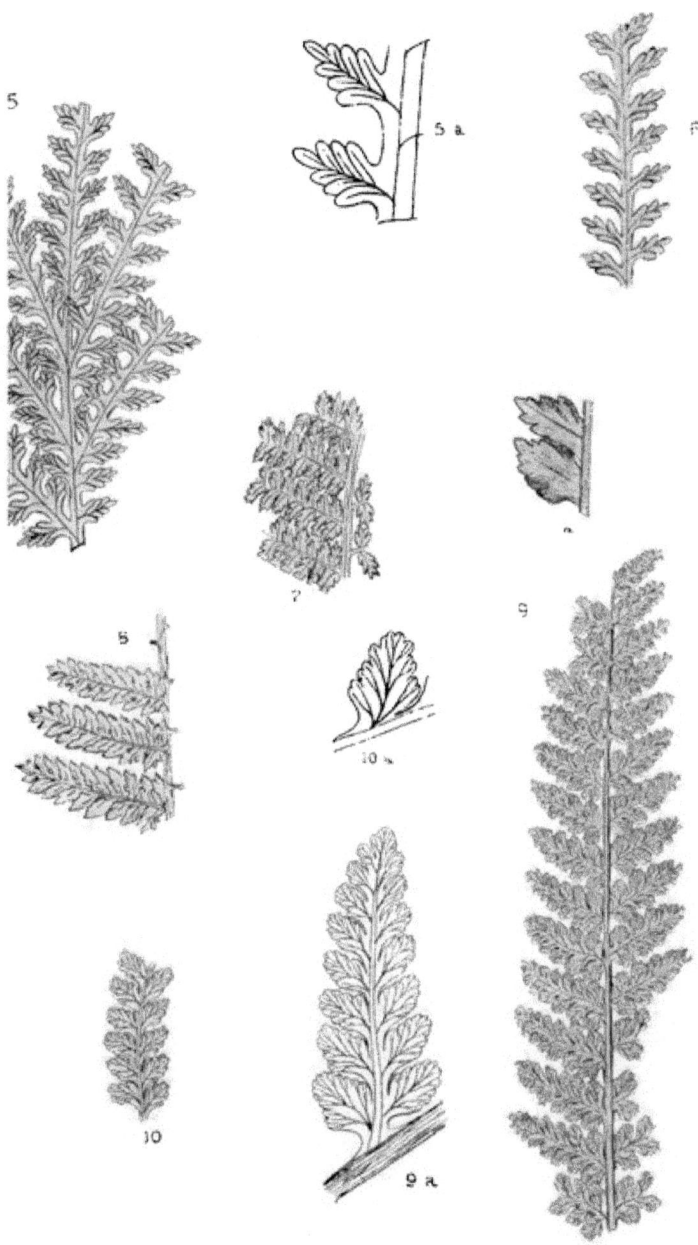

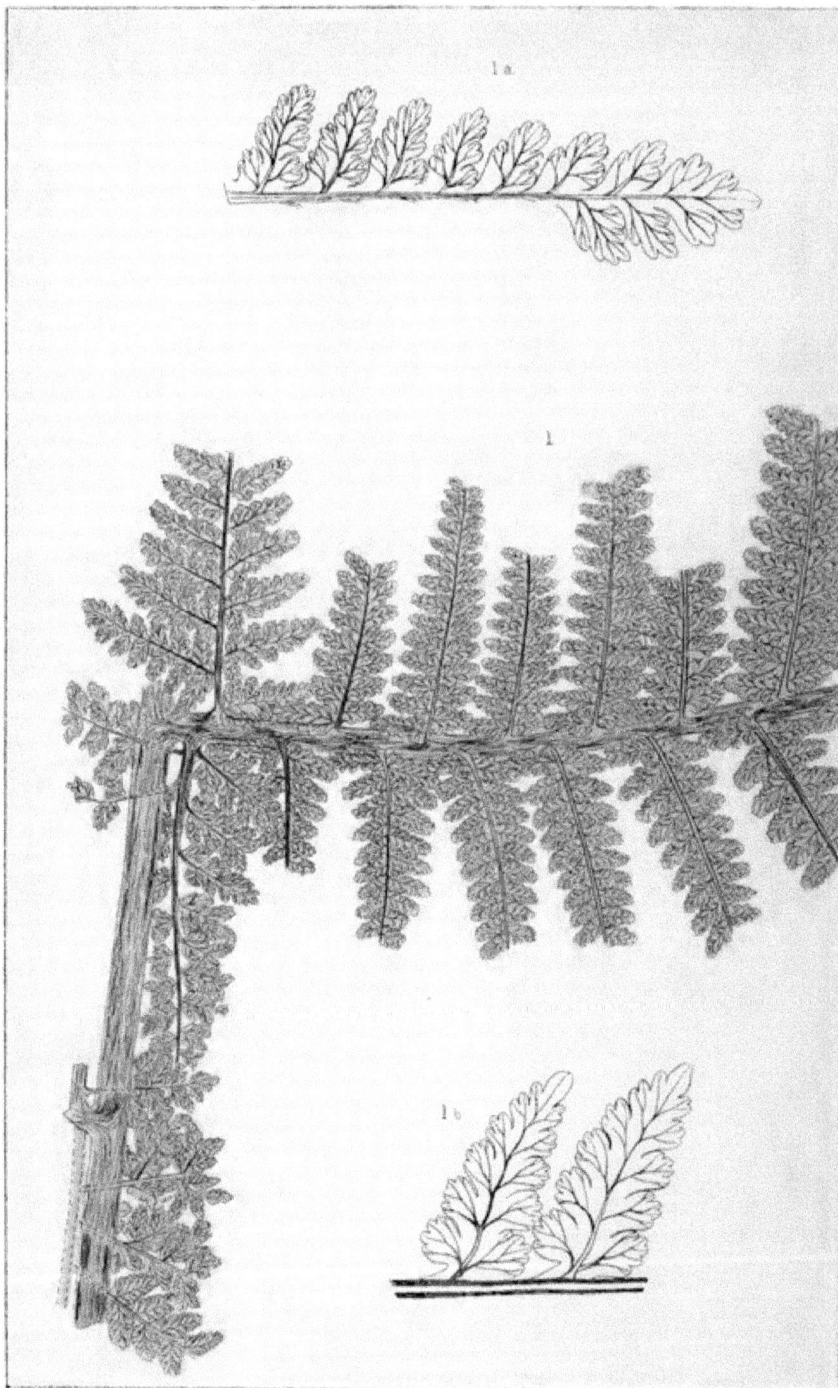

4 a.

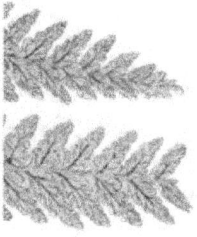

5 a.

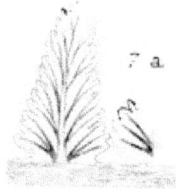

7 a.

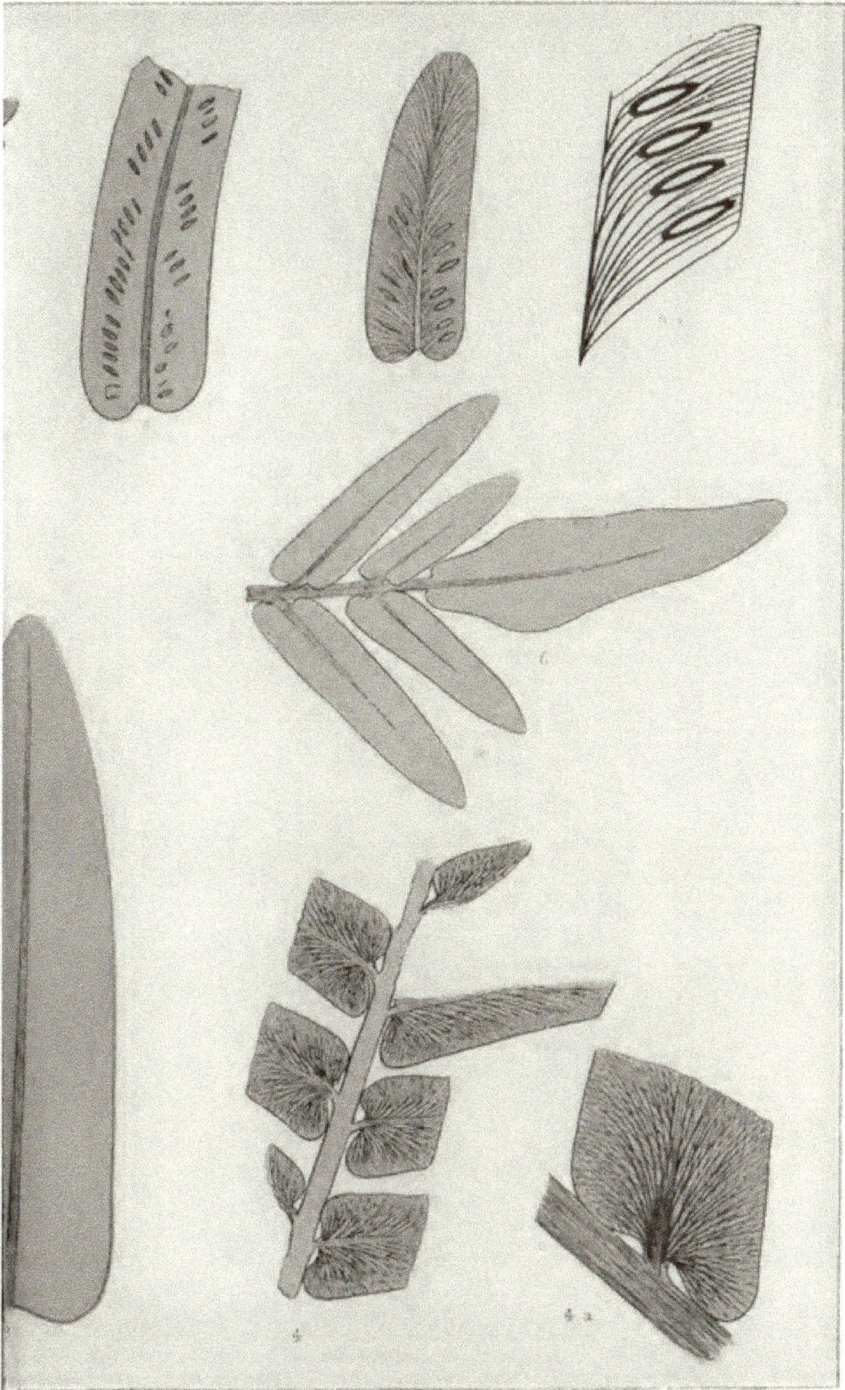

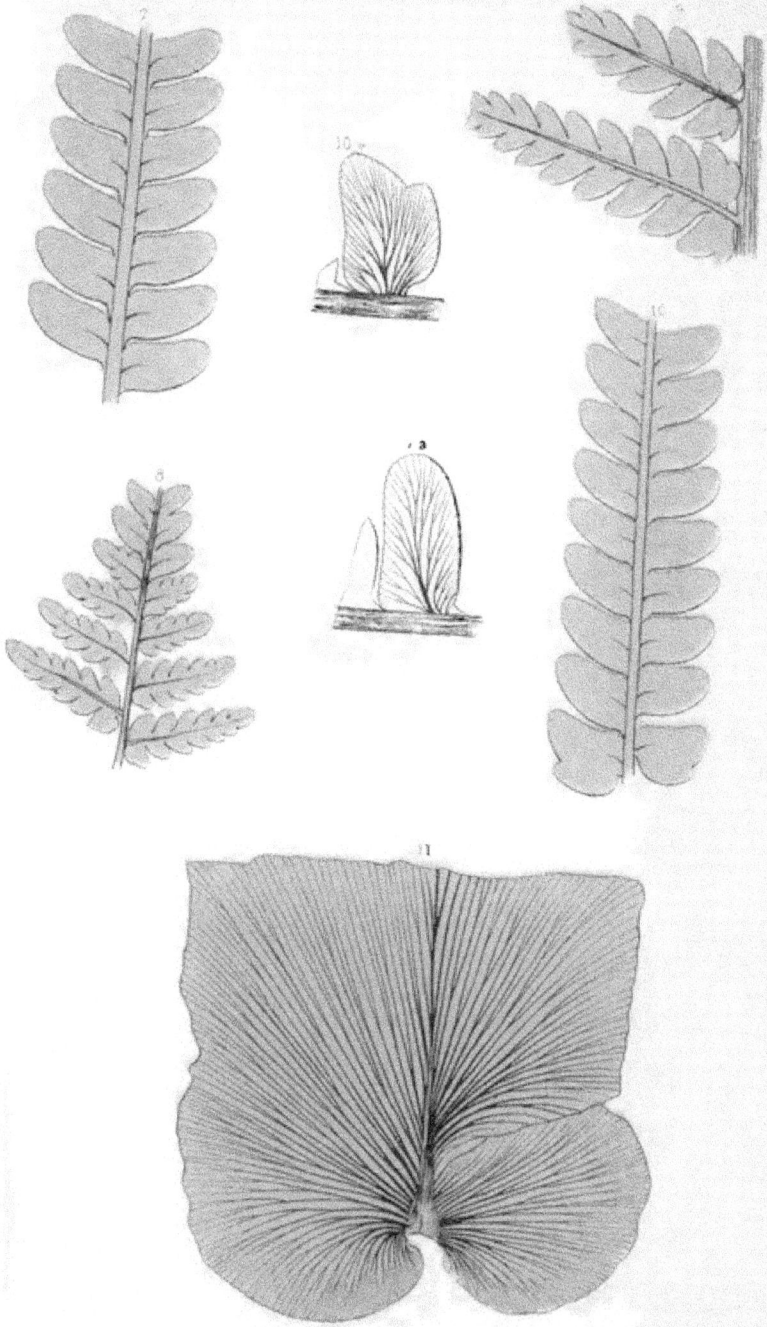

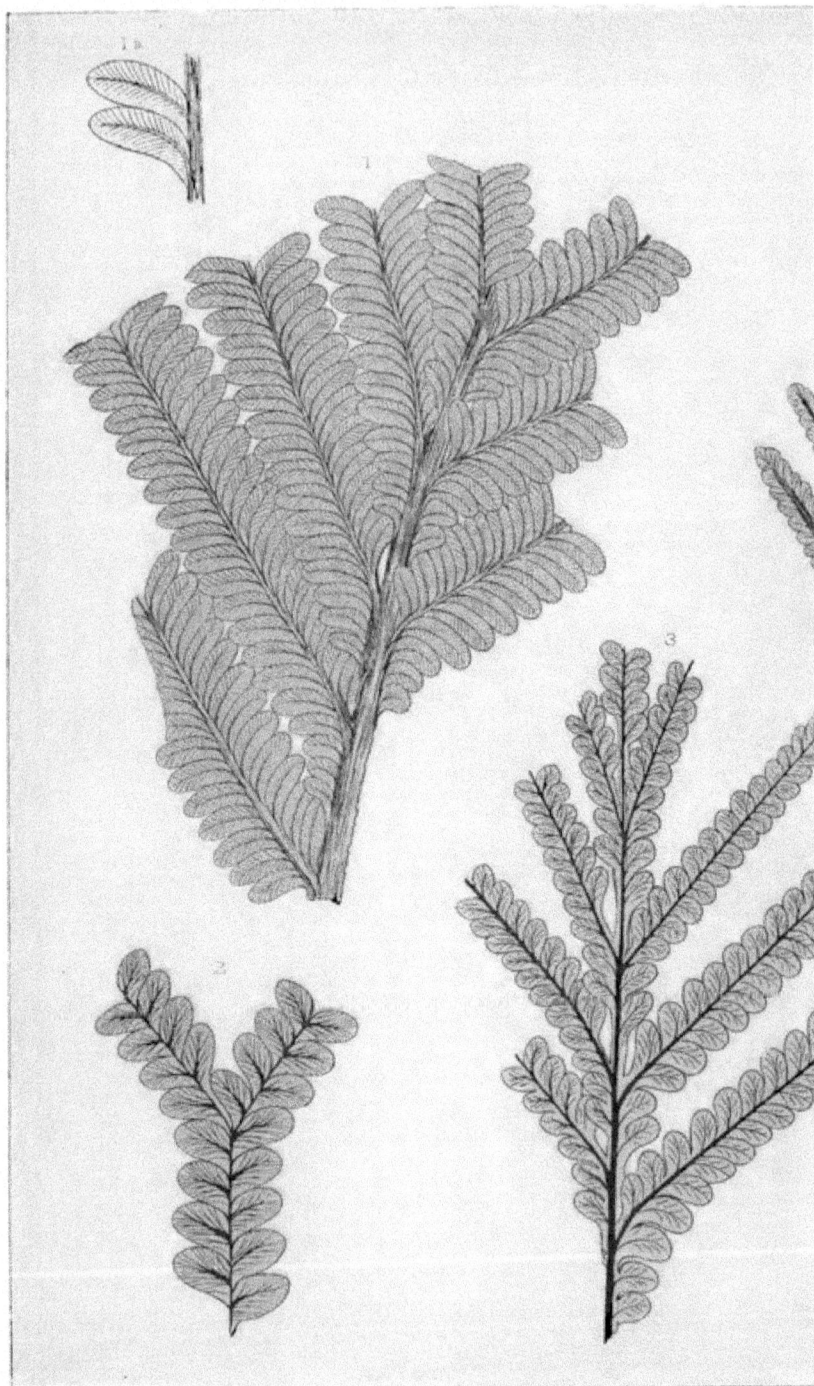

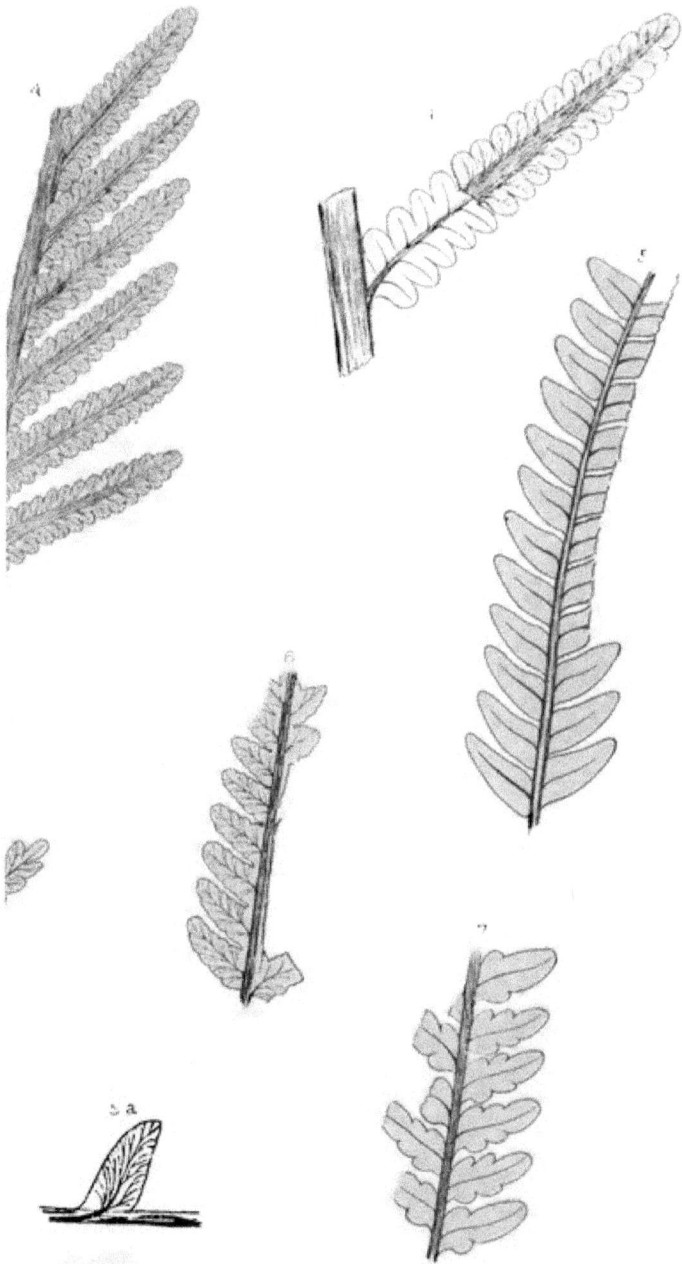

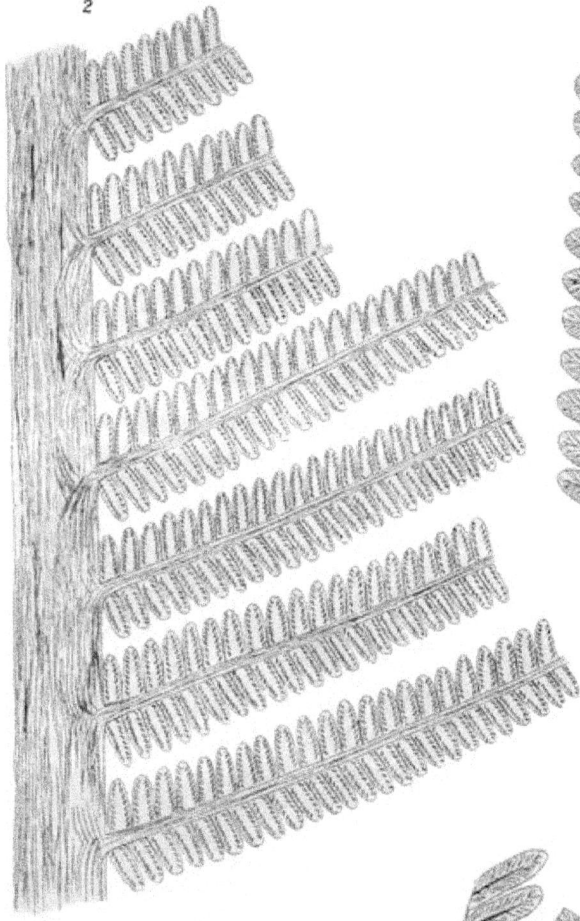

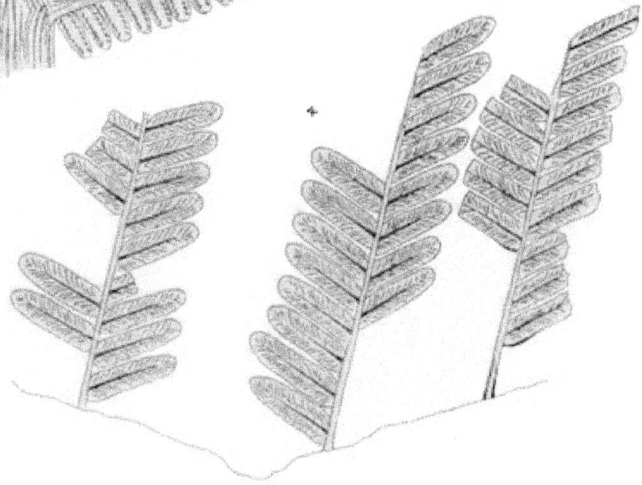

1a

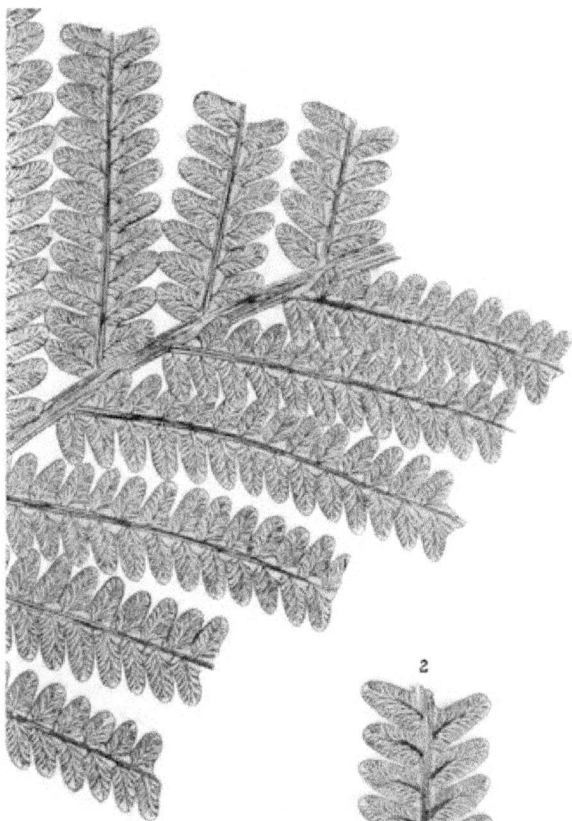

2

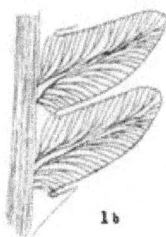

1 b

Lane S. Hart, State Printer

1

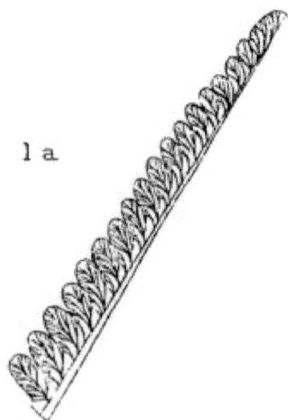

1 a

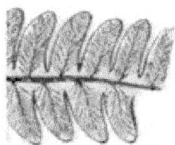

2 a.

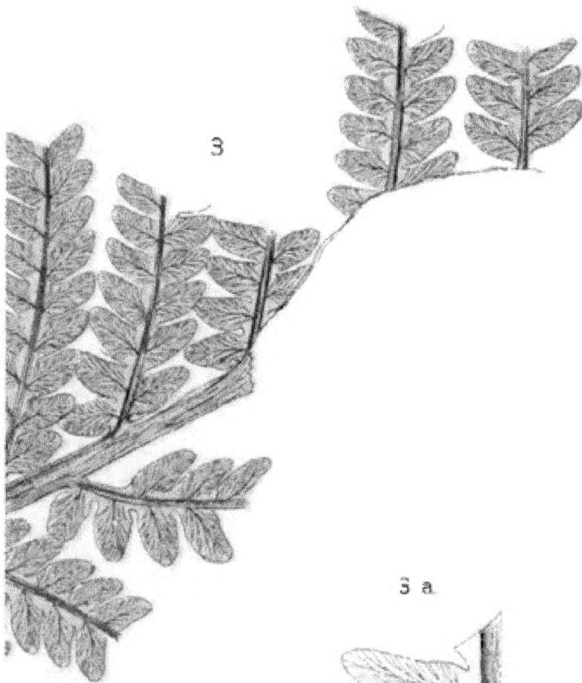

3

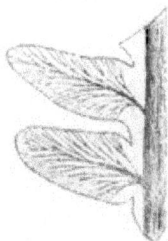

3 a.

1

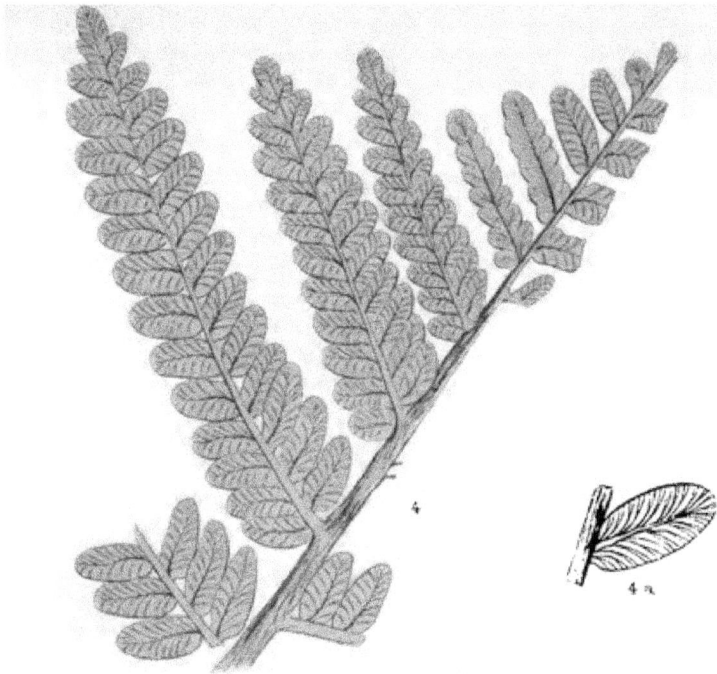

4

4 a

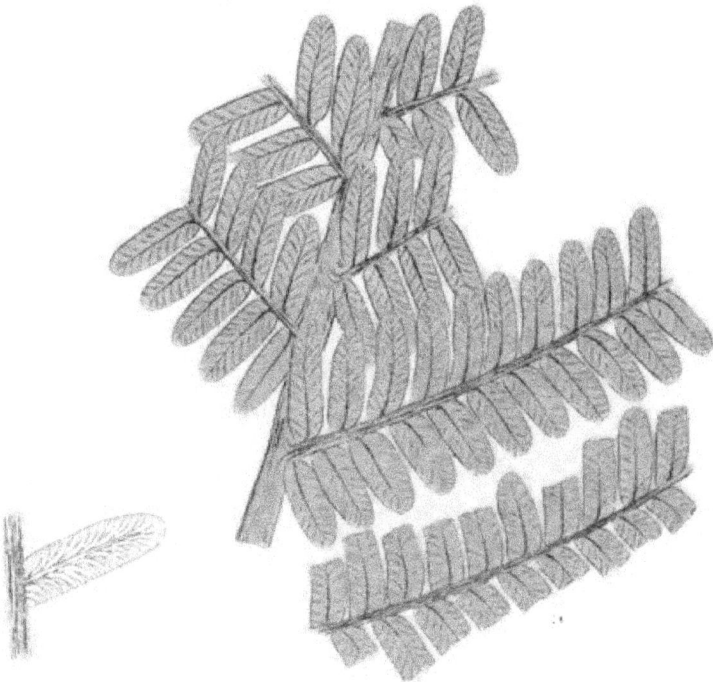

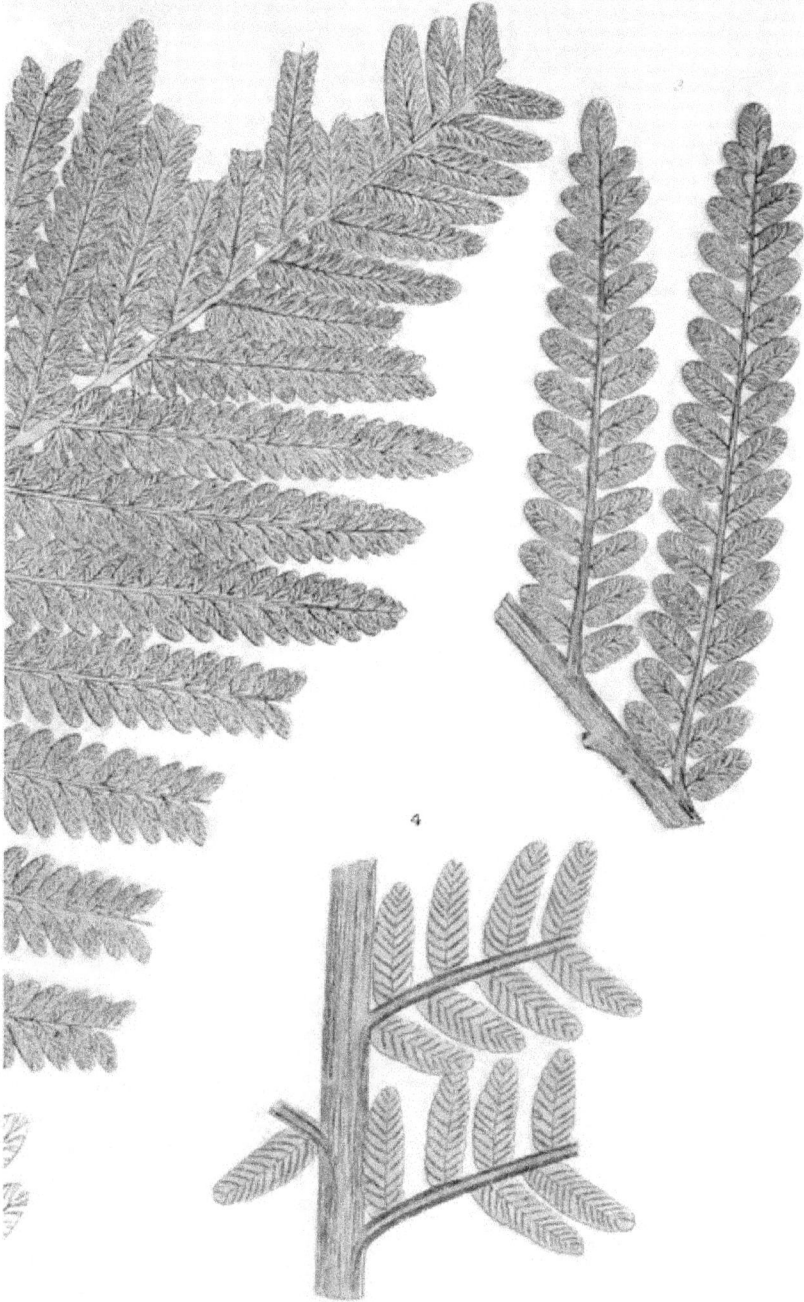

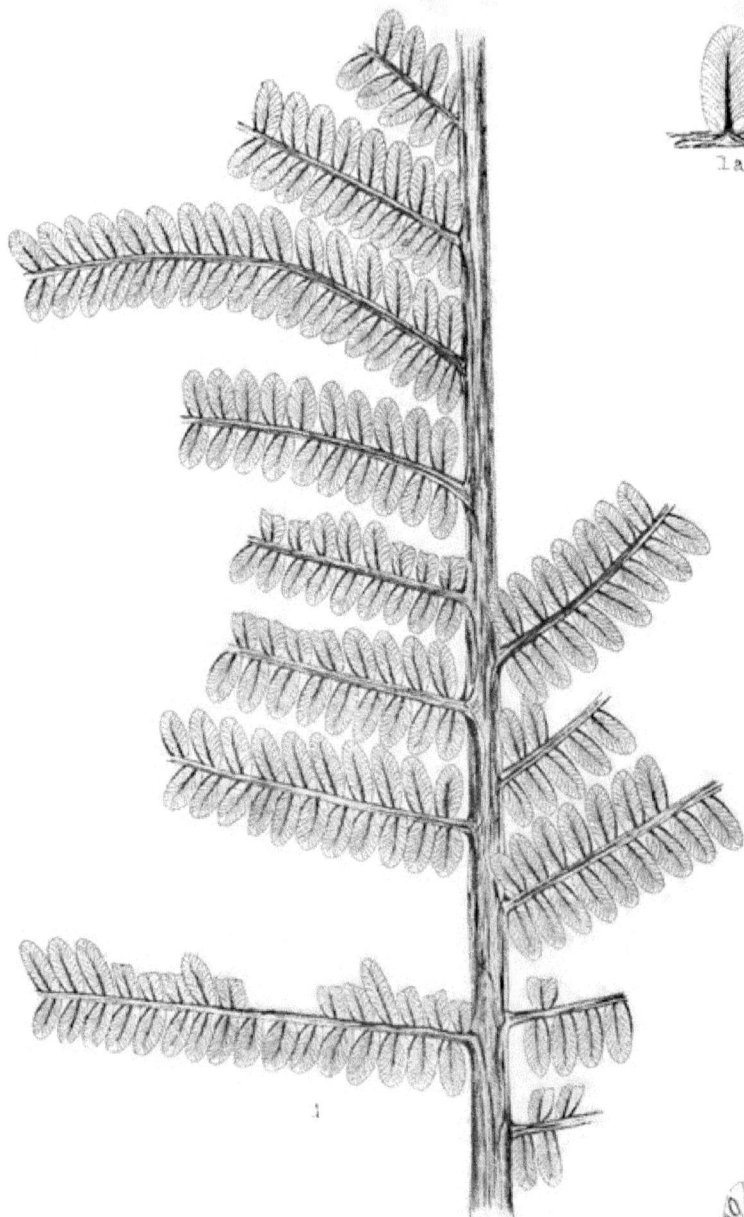

1a

1

6a

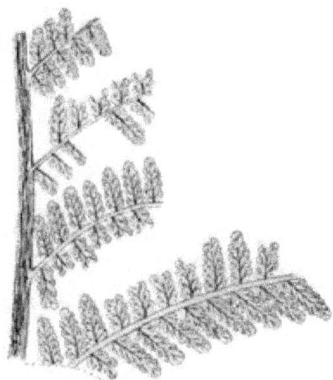

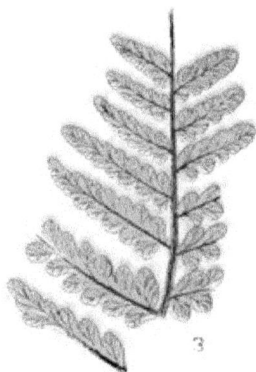

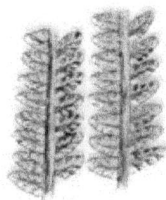

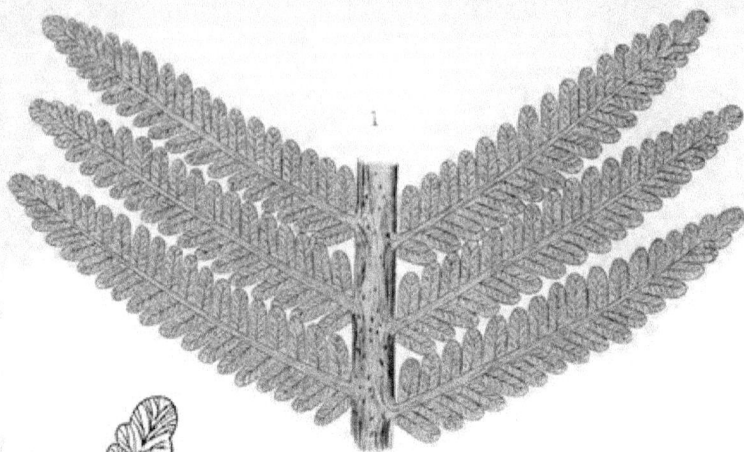

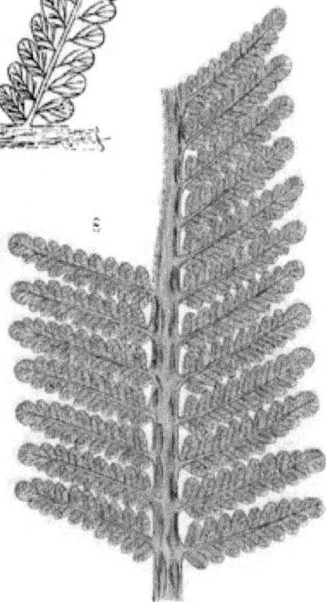

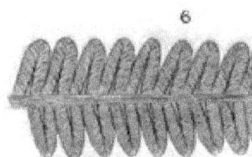

Lane S. Hart, State Printer.

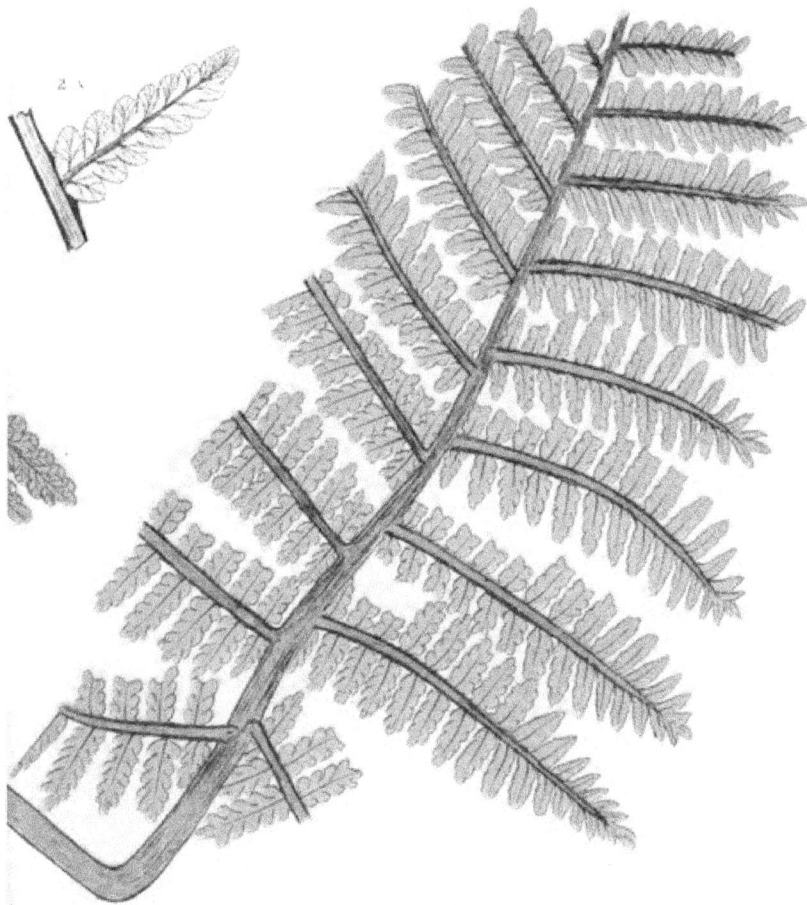

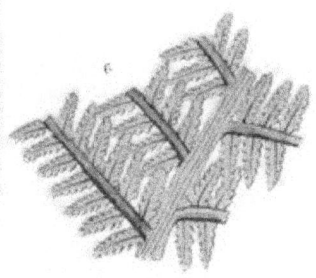

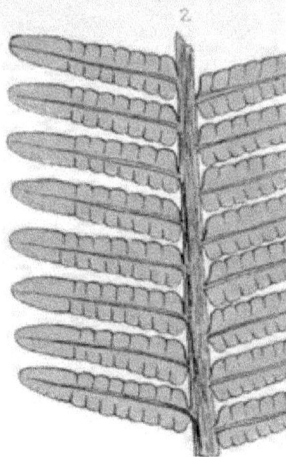
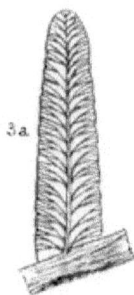
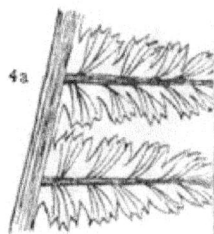
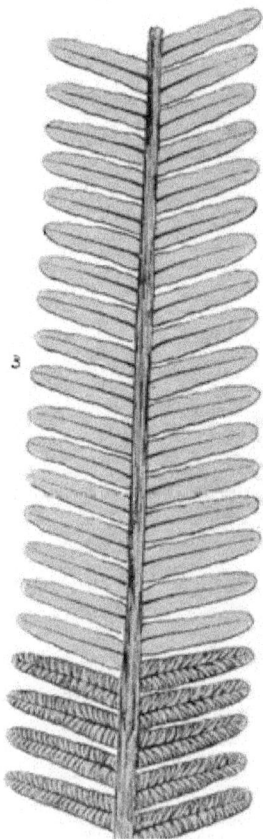

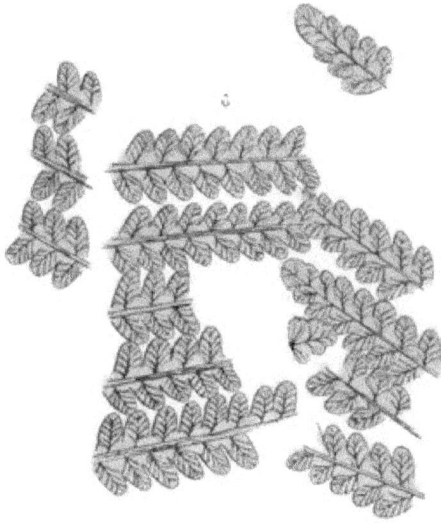

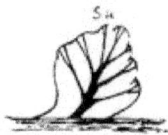

1 b

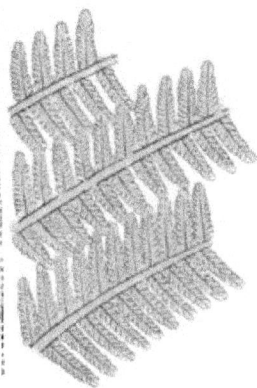

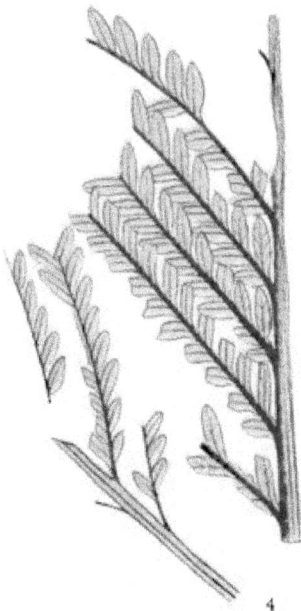

4

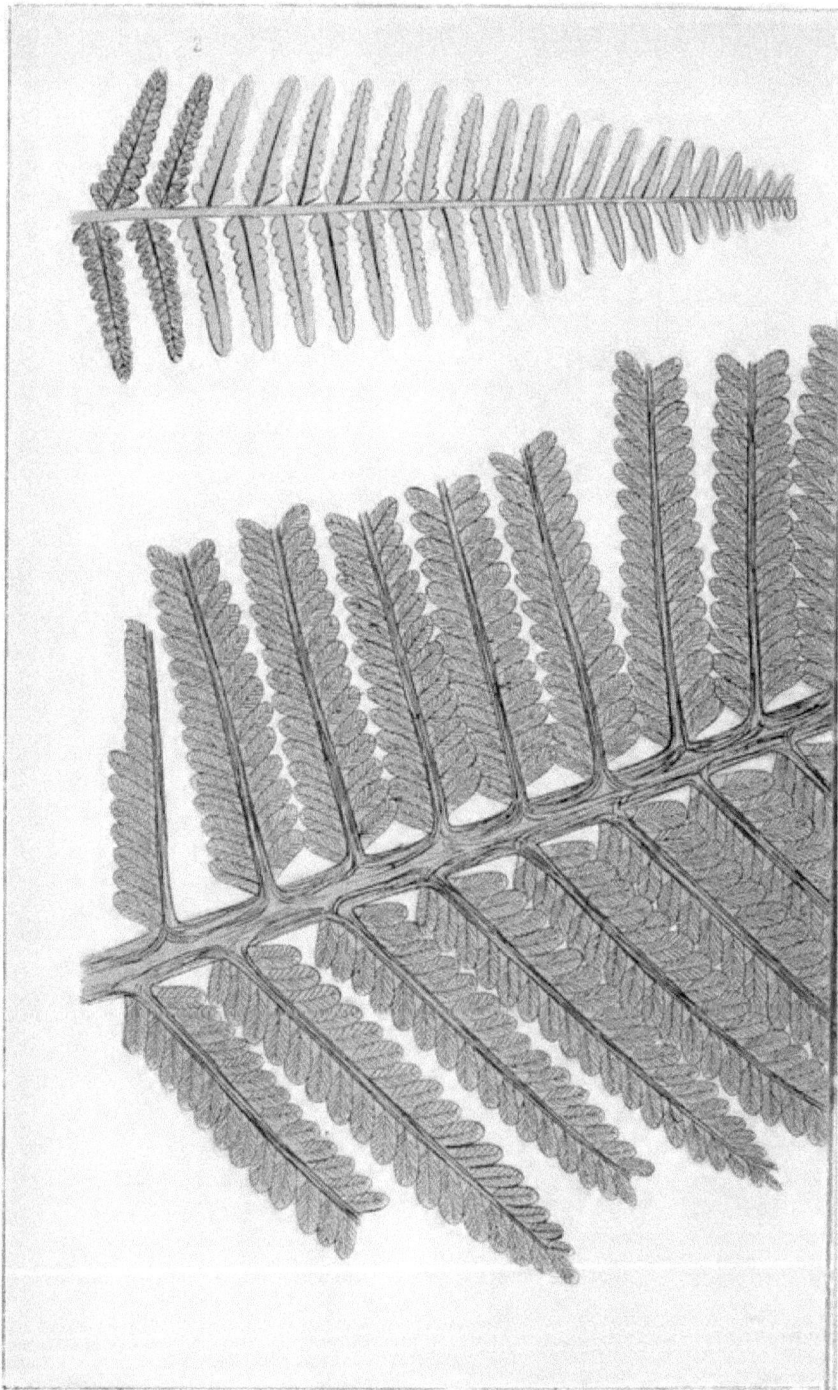

1a

1

7 a

7

2

4

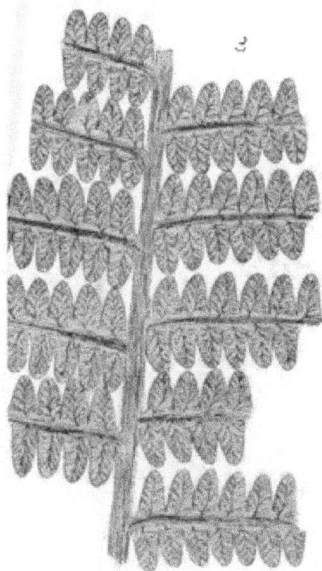

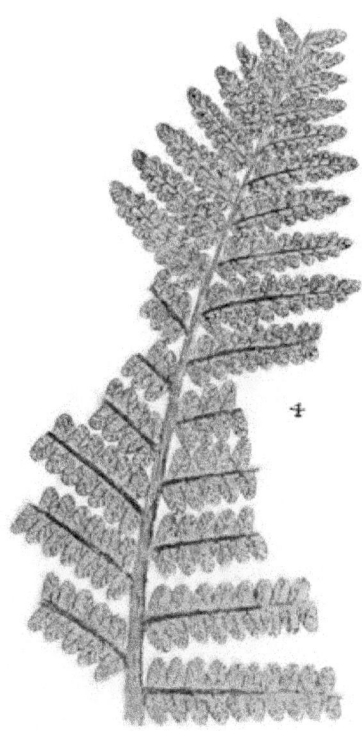

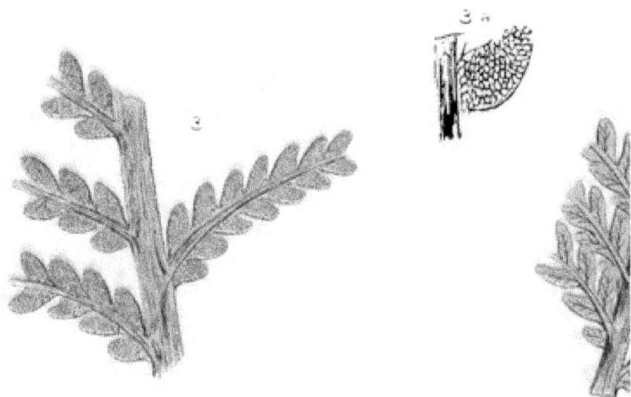

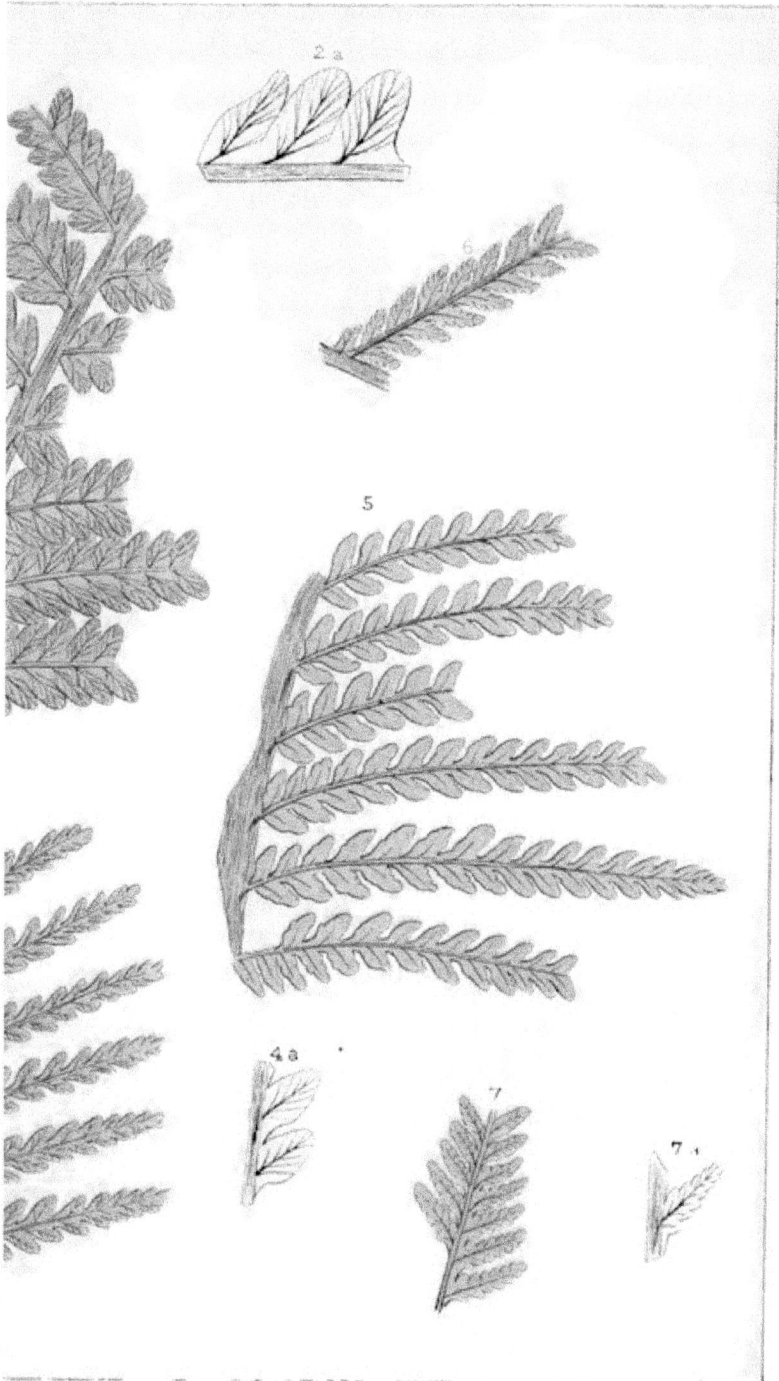

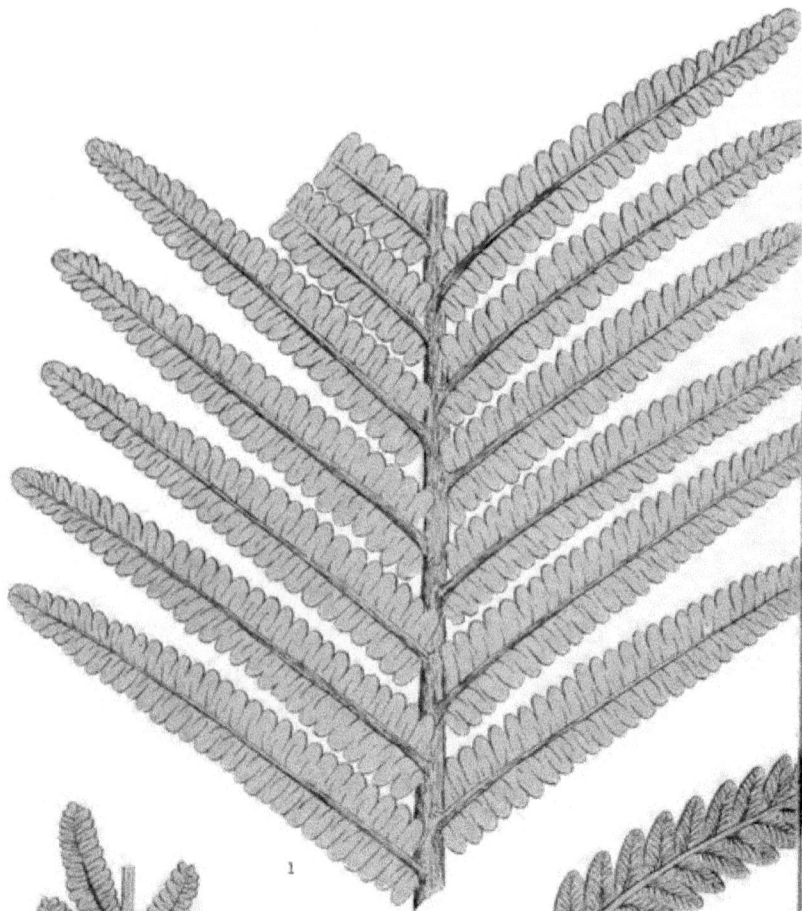

1

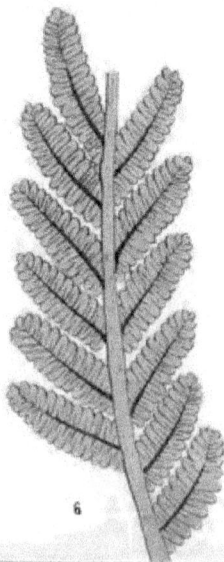

4

5 a

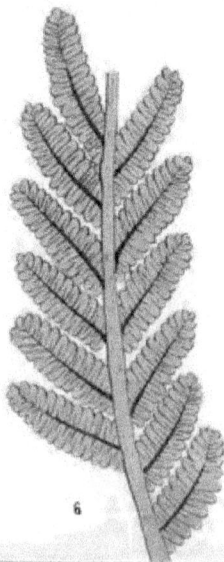

6

5

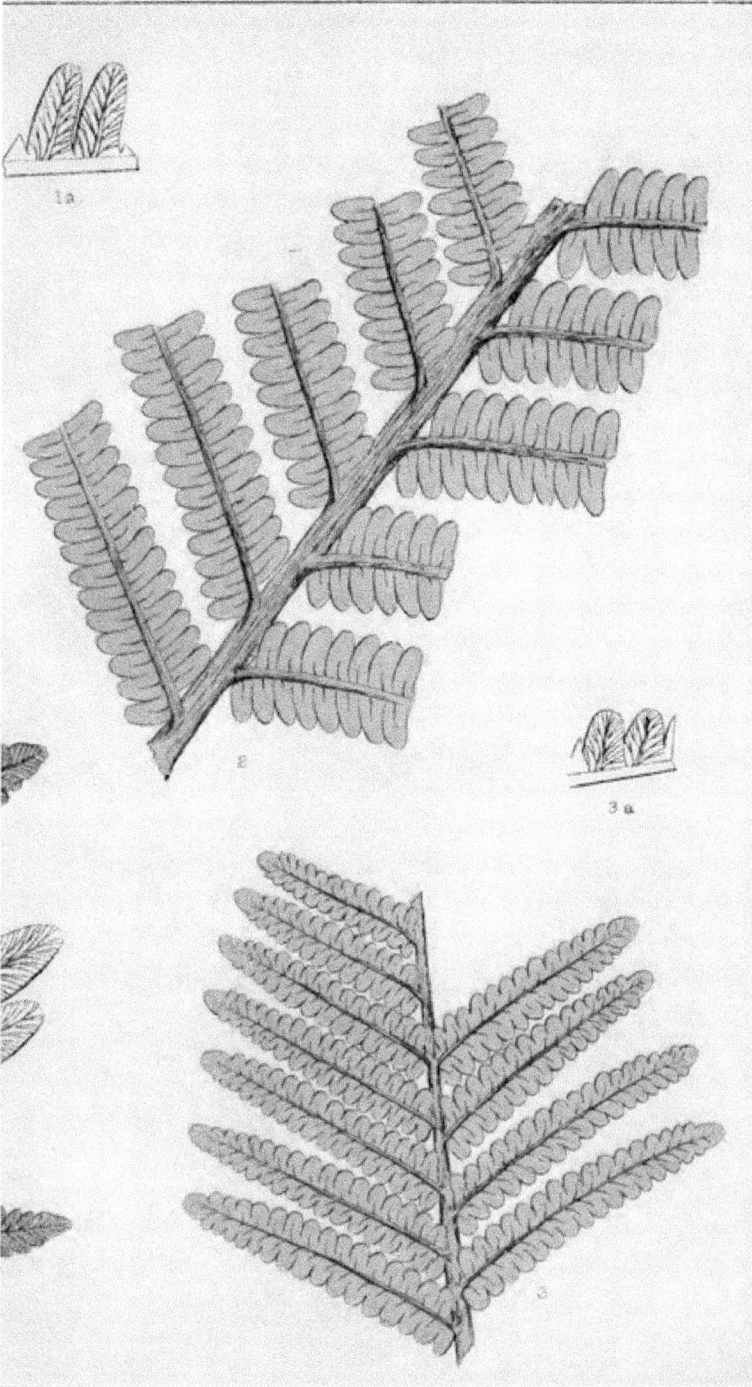

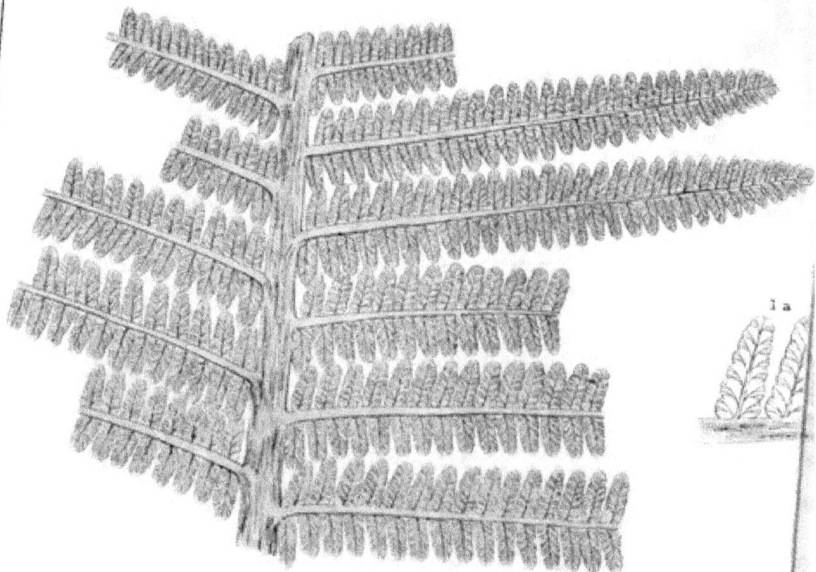

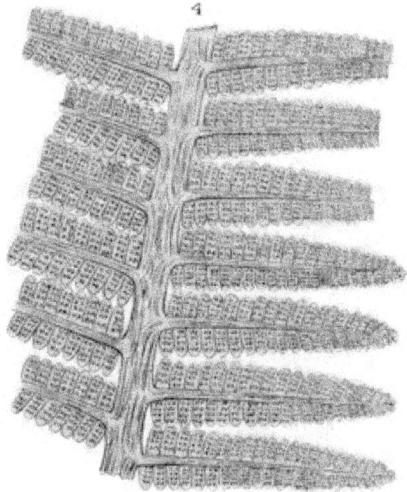

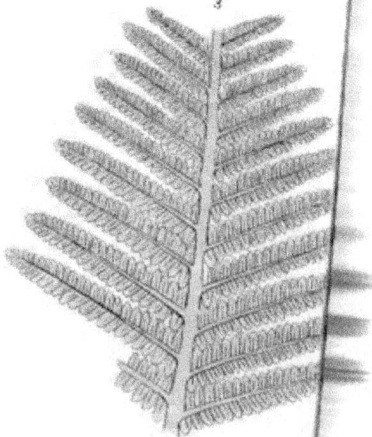

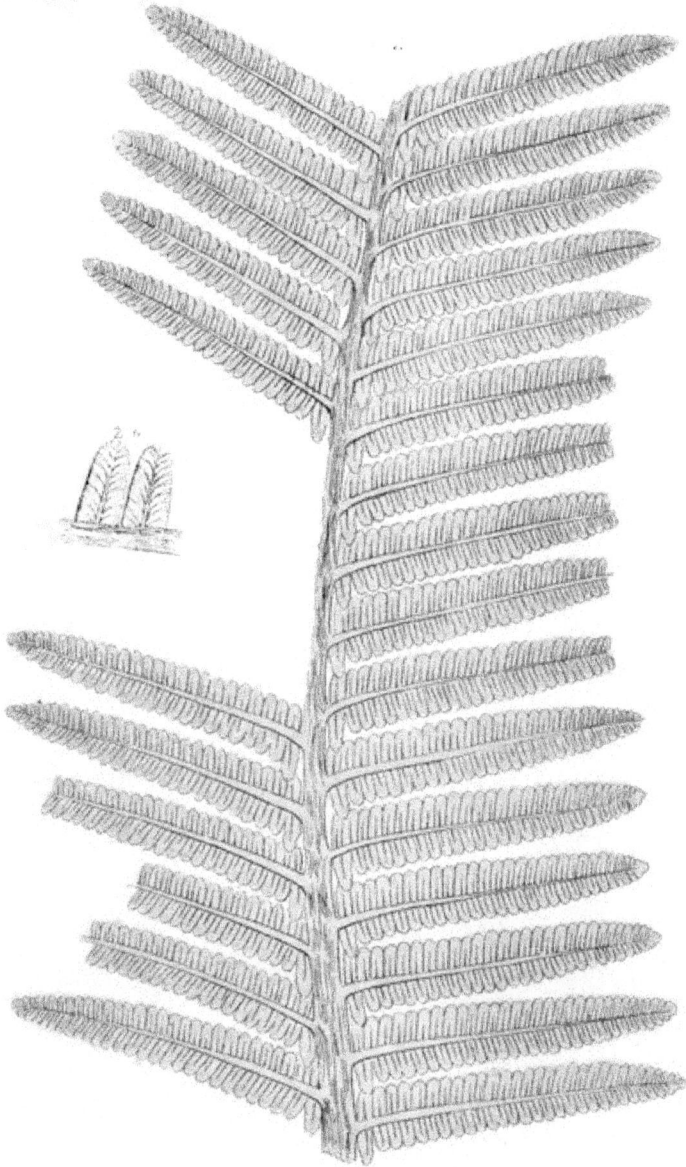

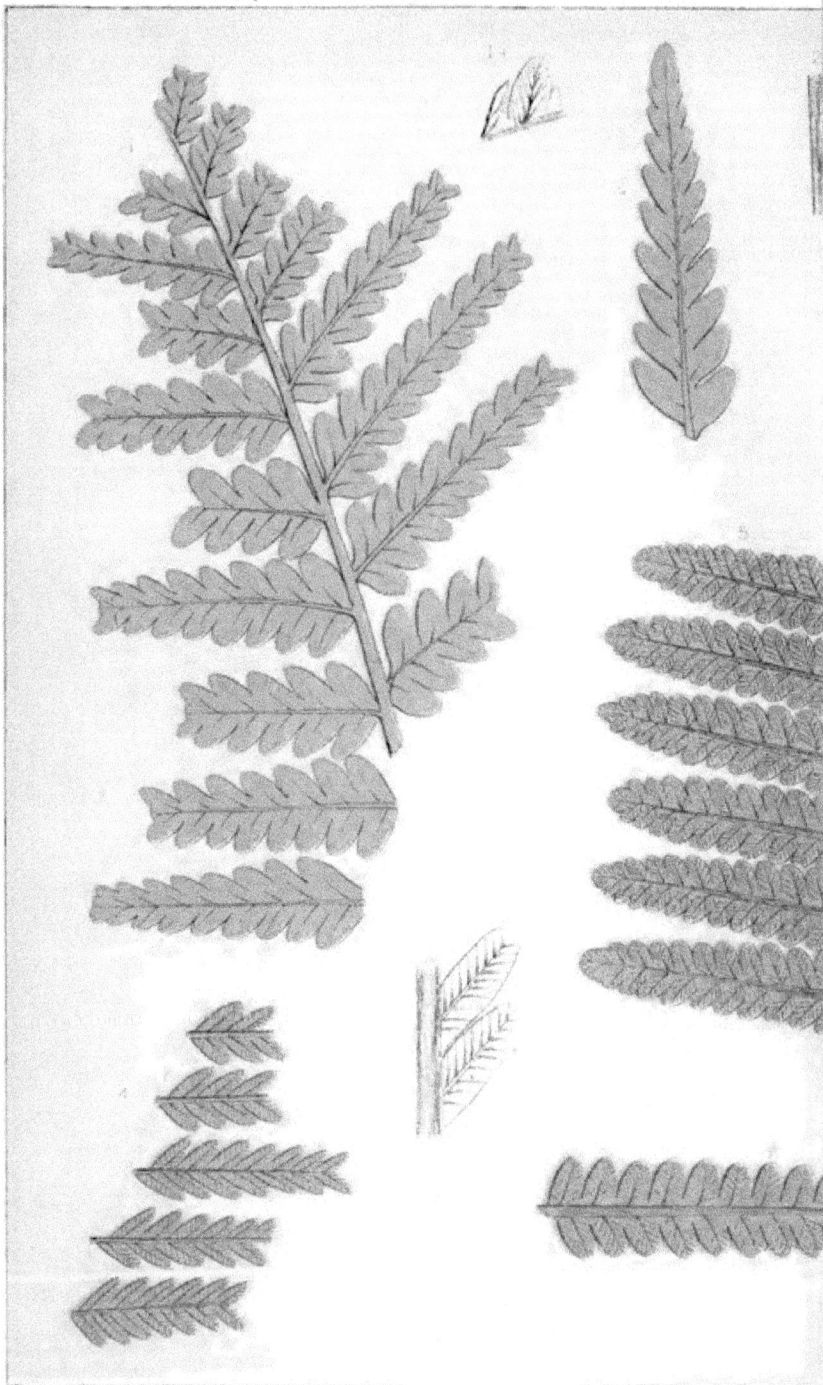

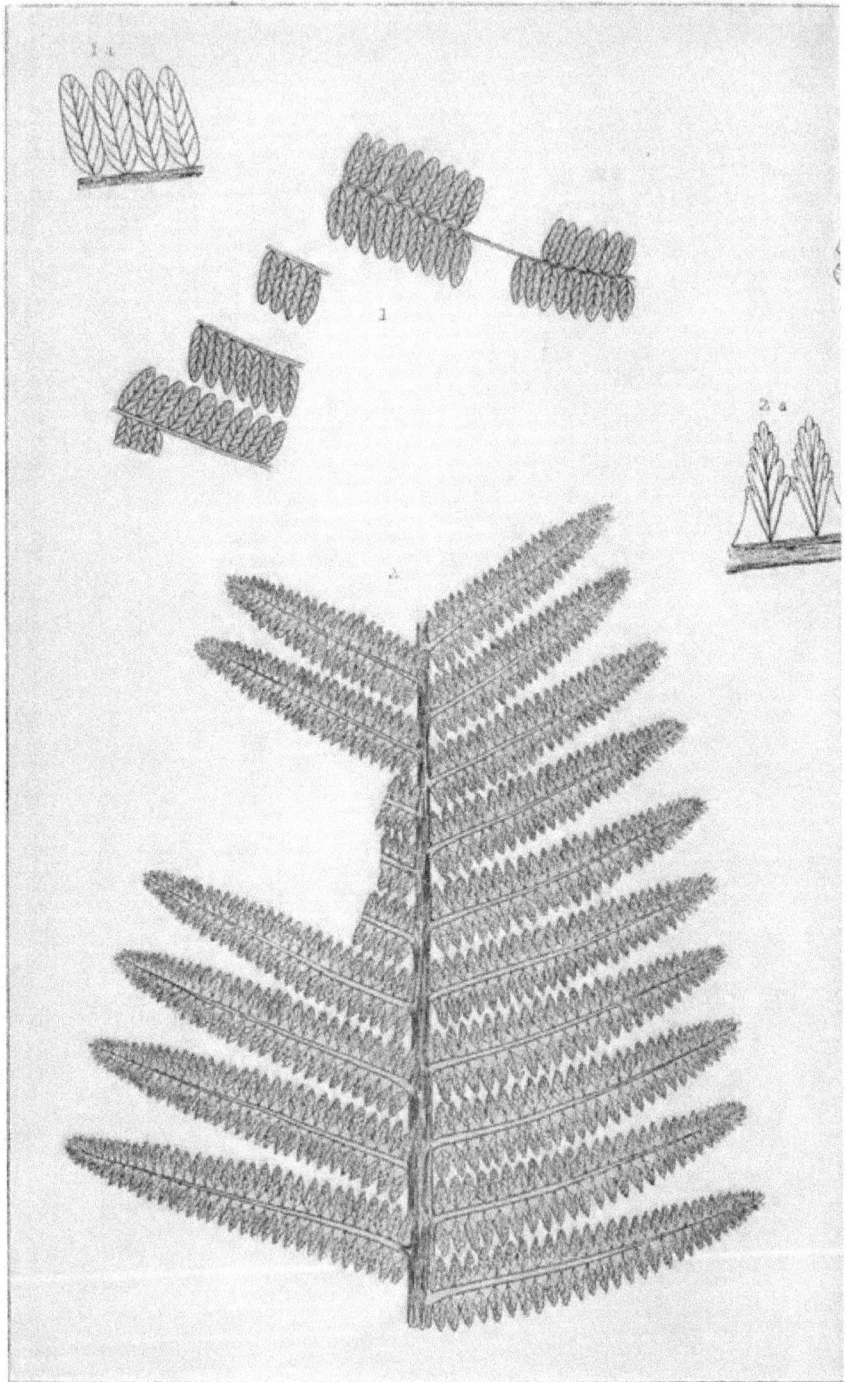

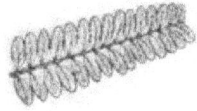

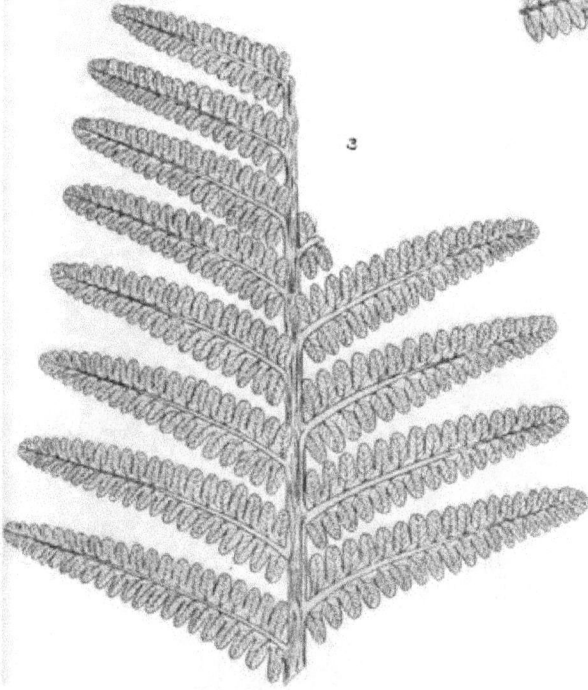

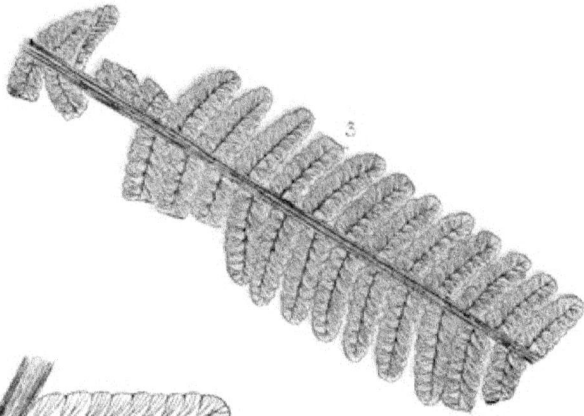

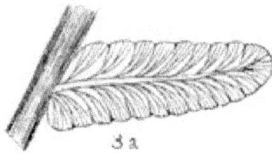

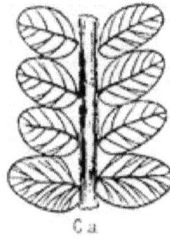

6 a

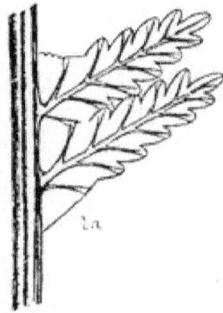

2 a

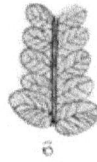

6

4 a

5

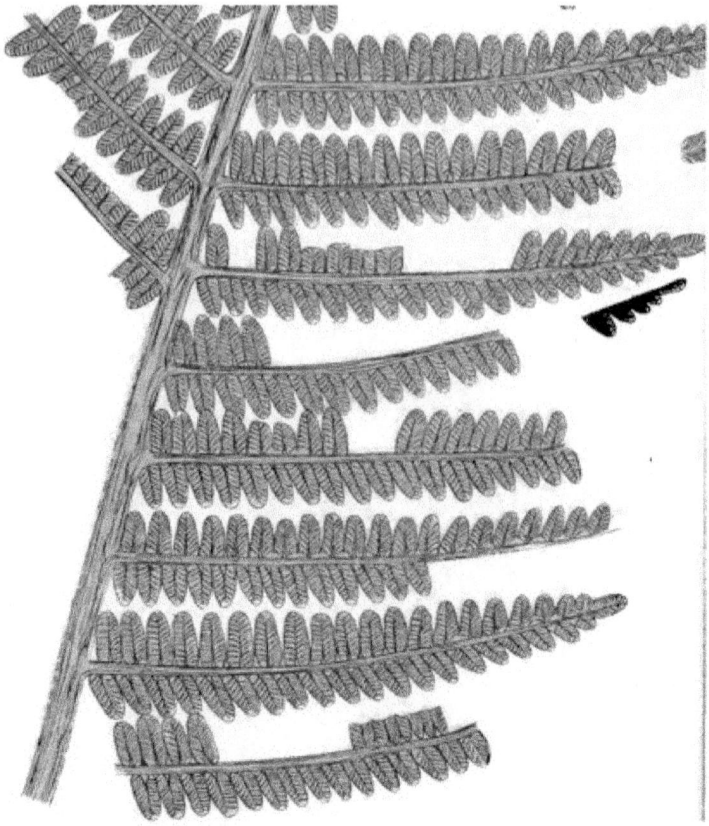

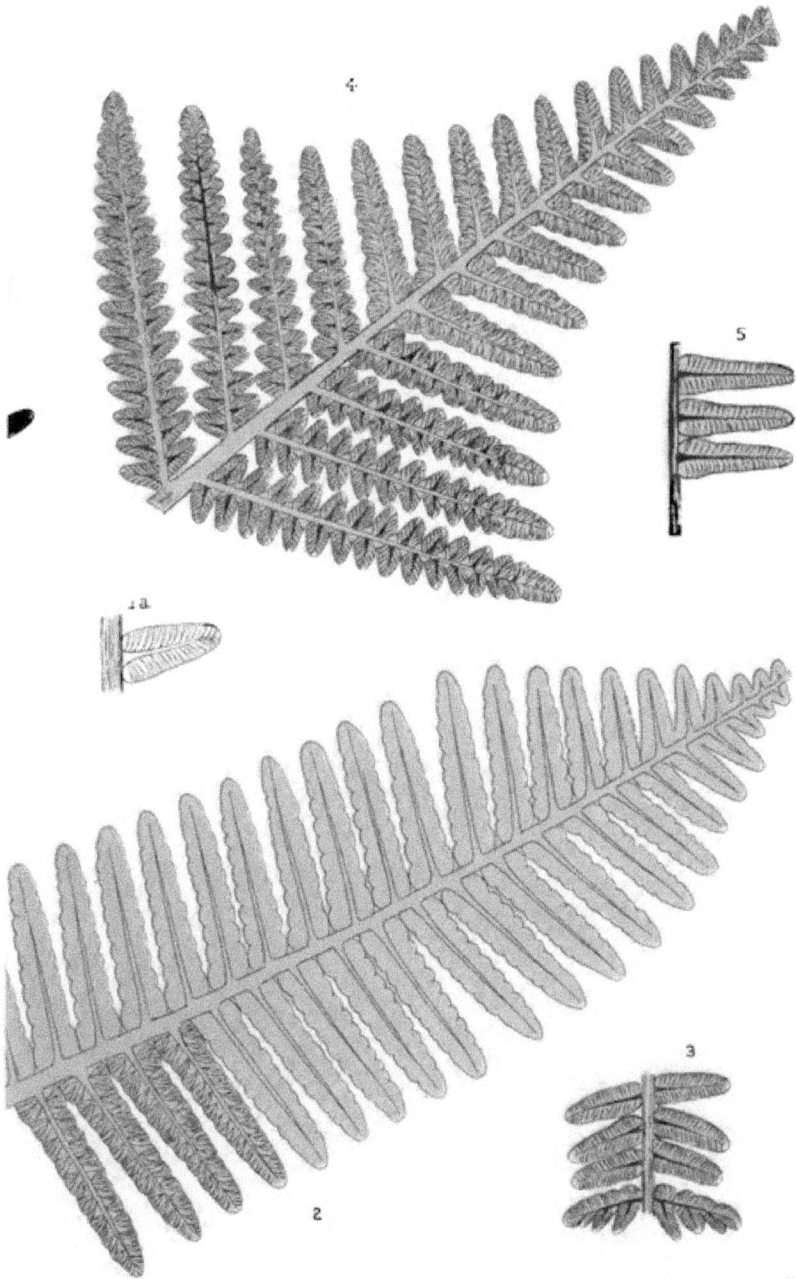

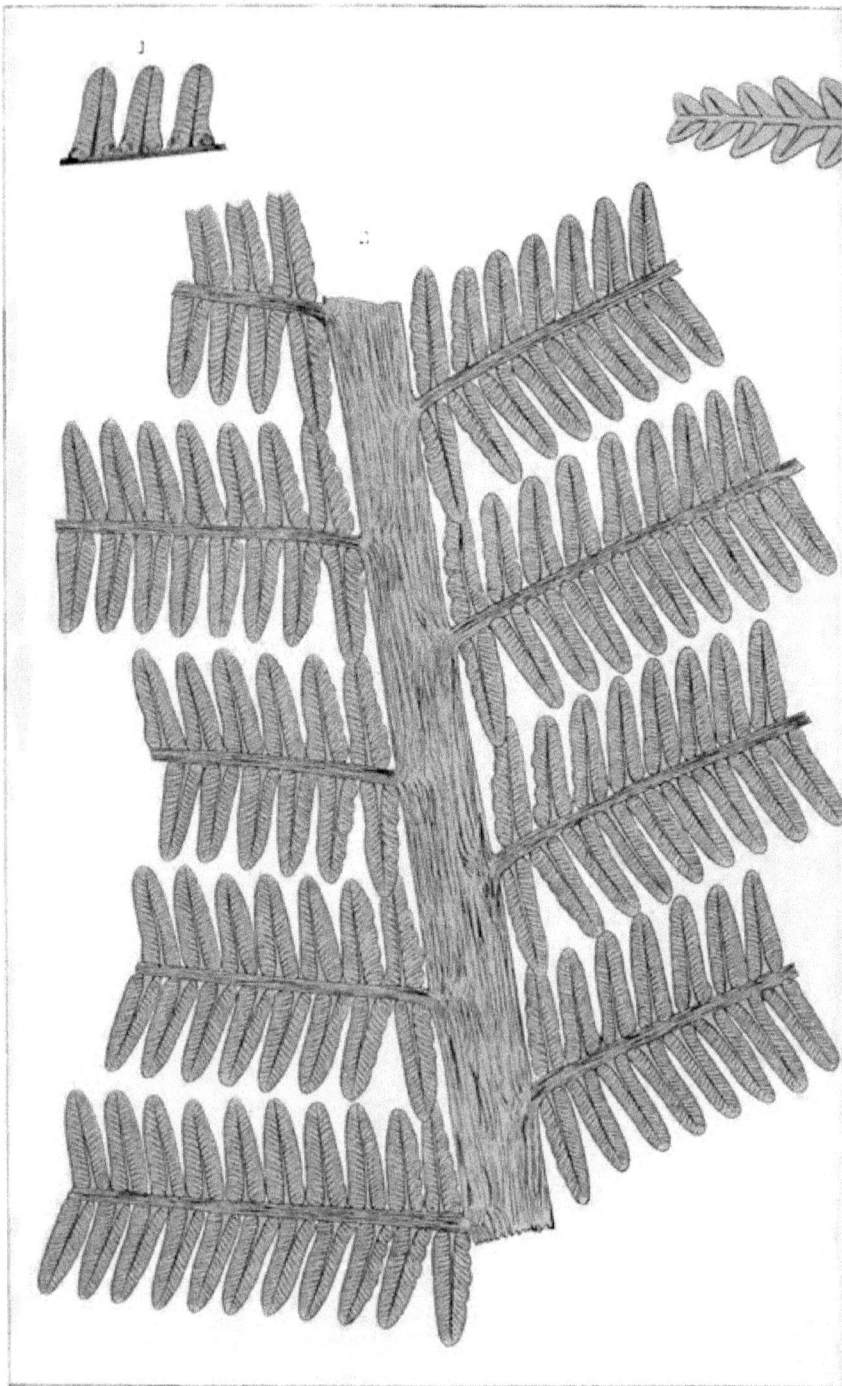

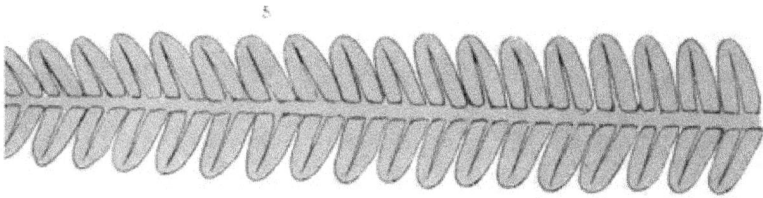

5

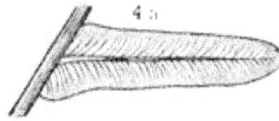

4 5

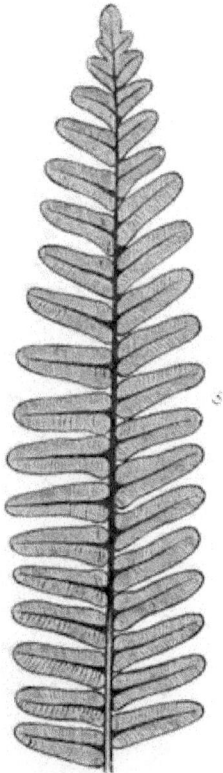

5

4

6

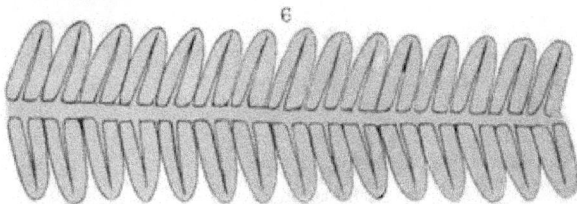

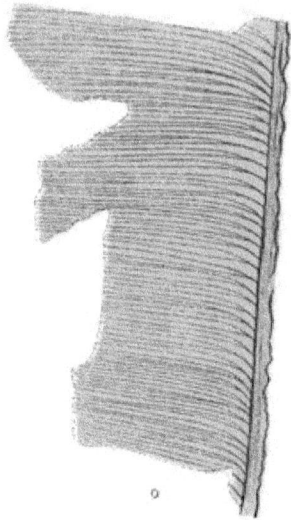

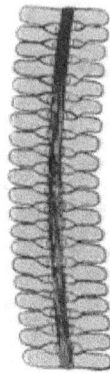

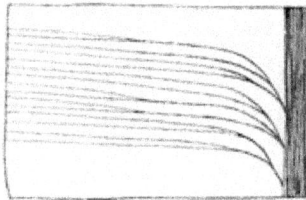

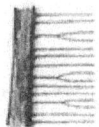

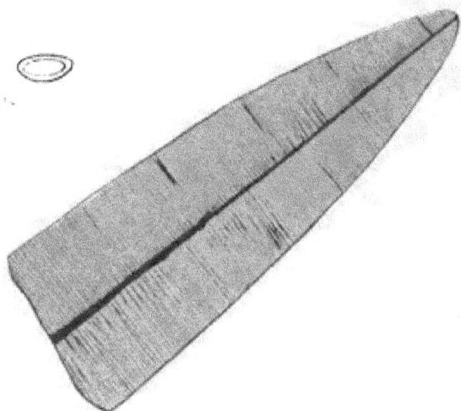

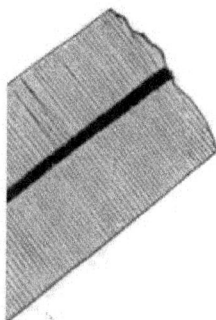

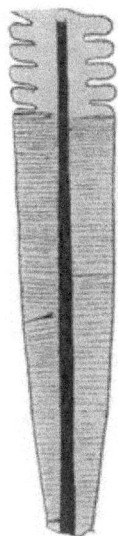

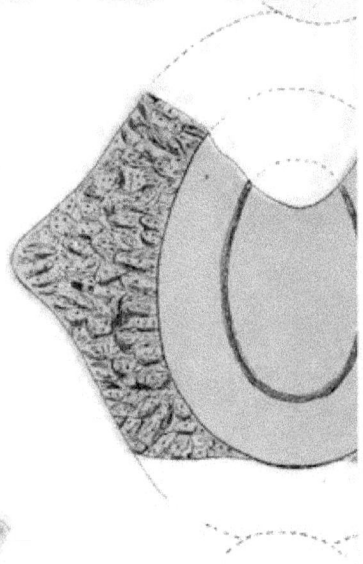

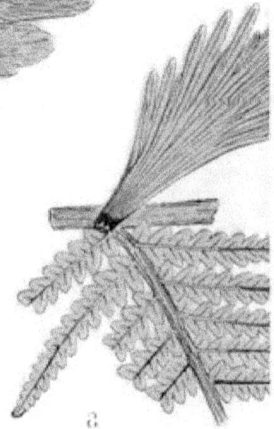

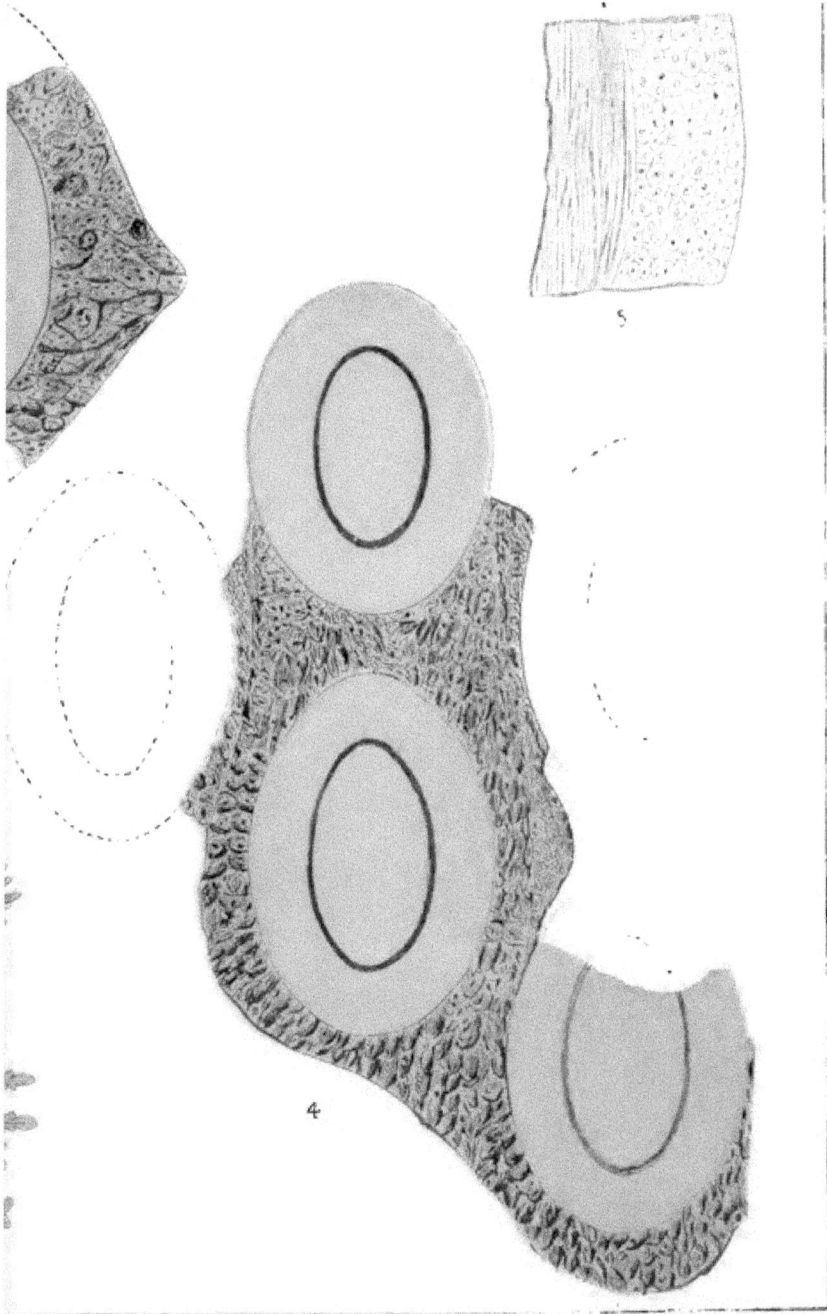

5

4

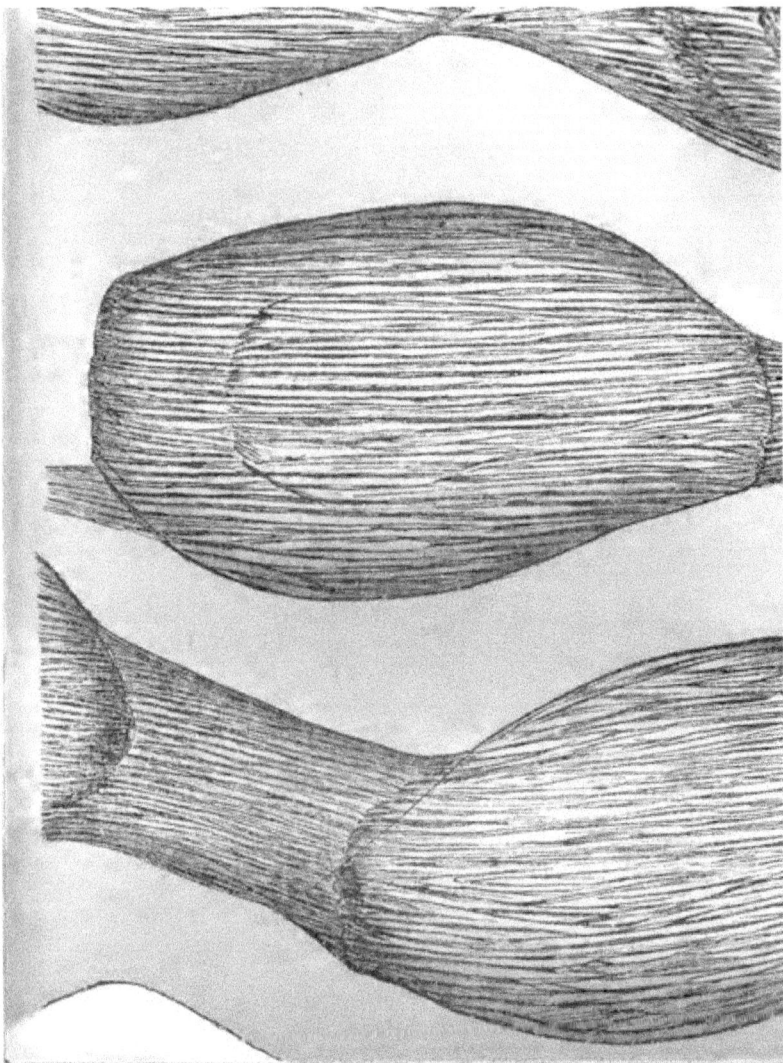

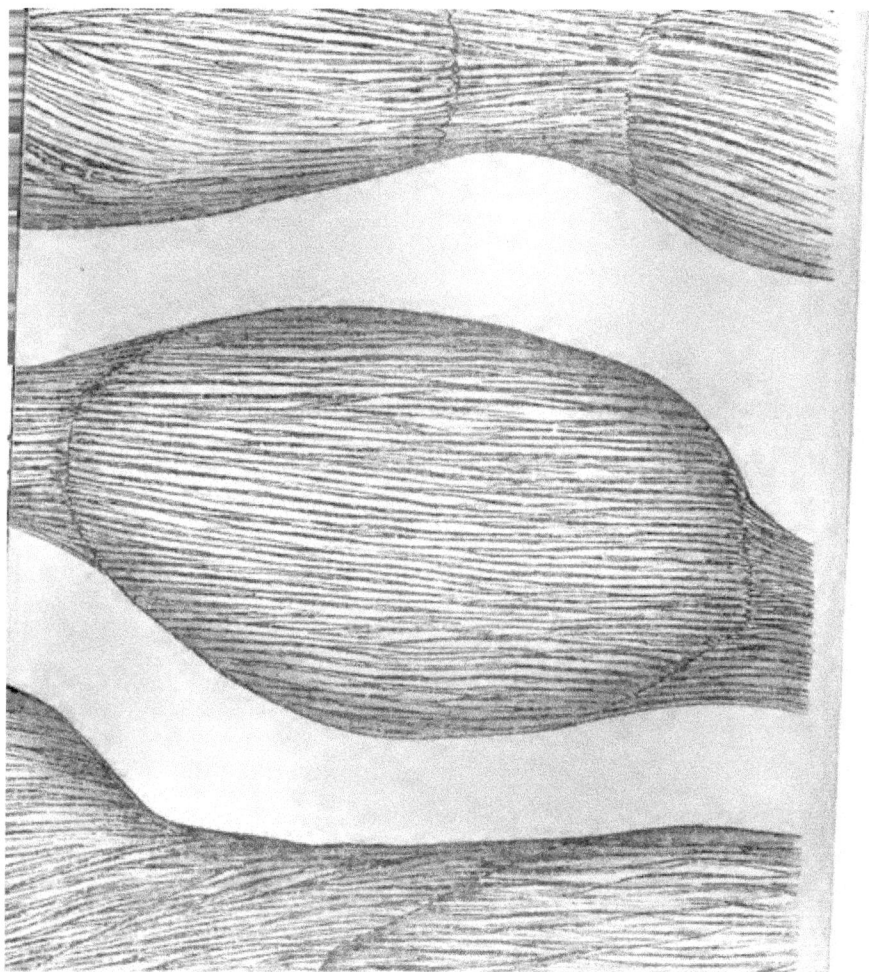

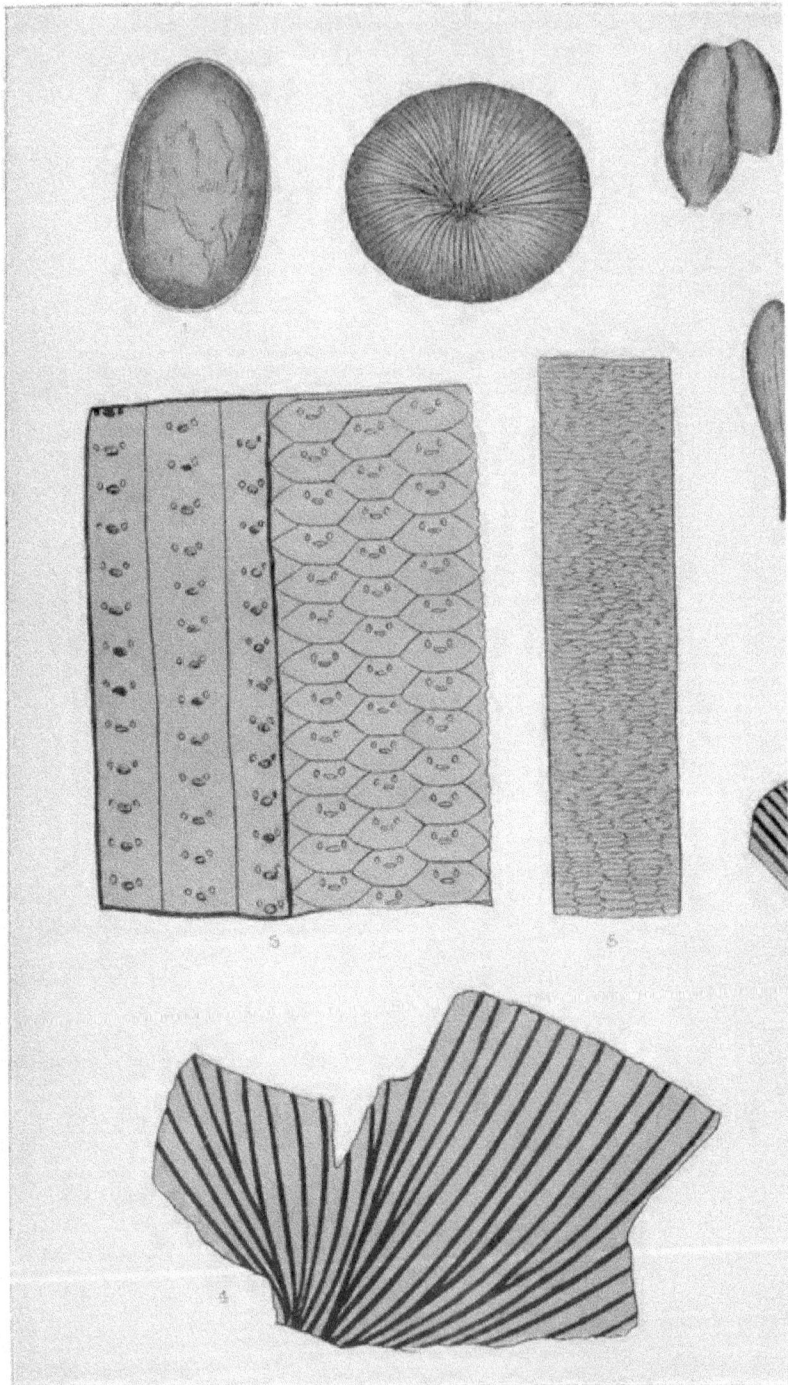

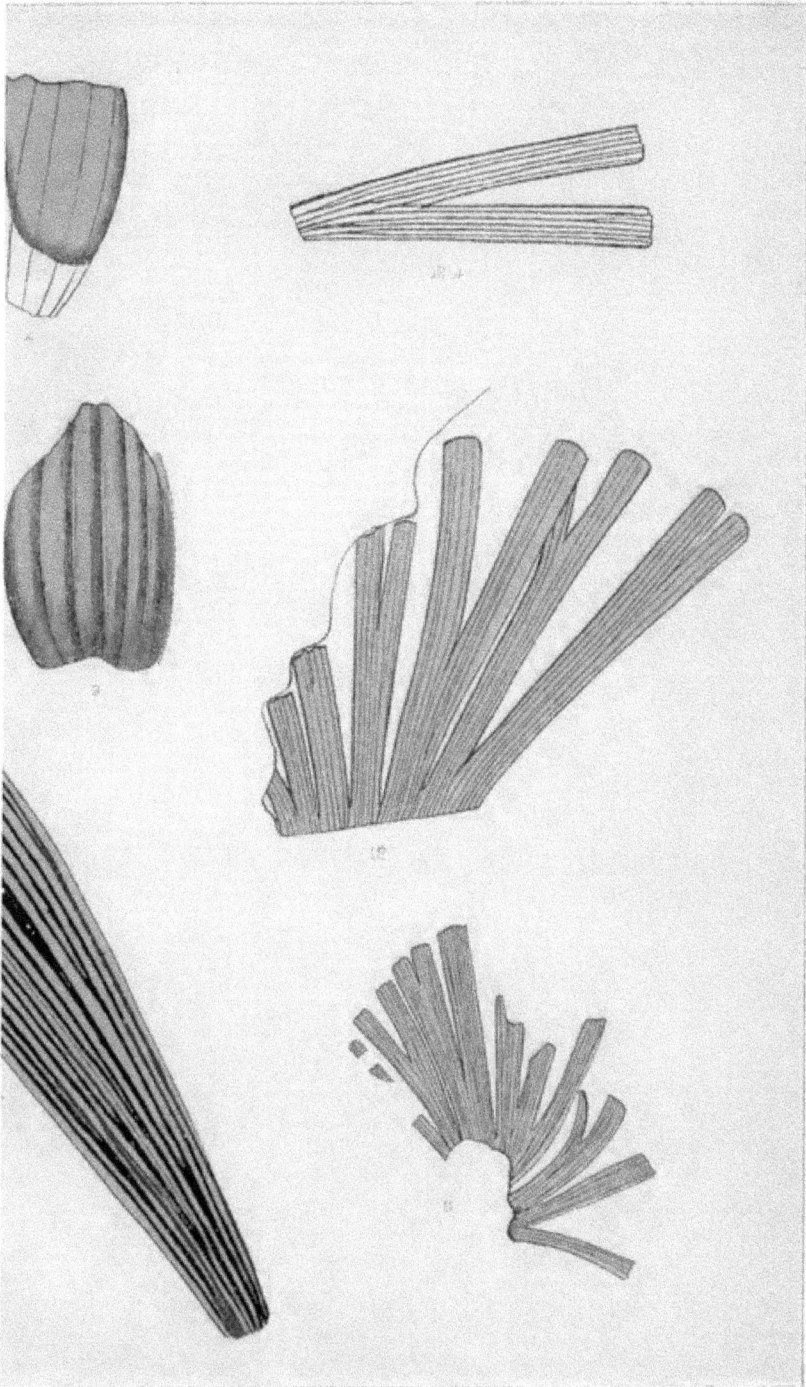

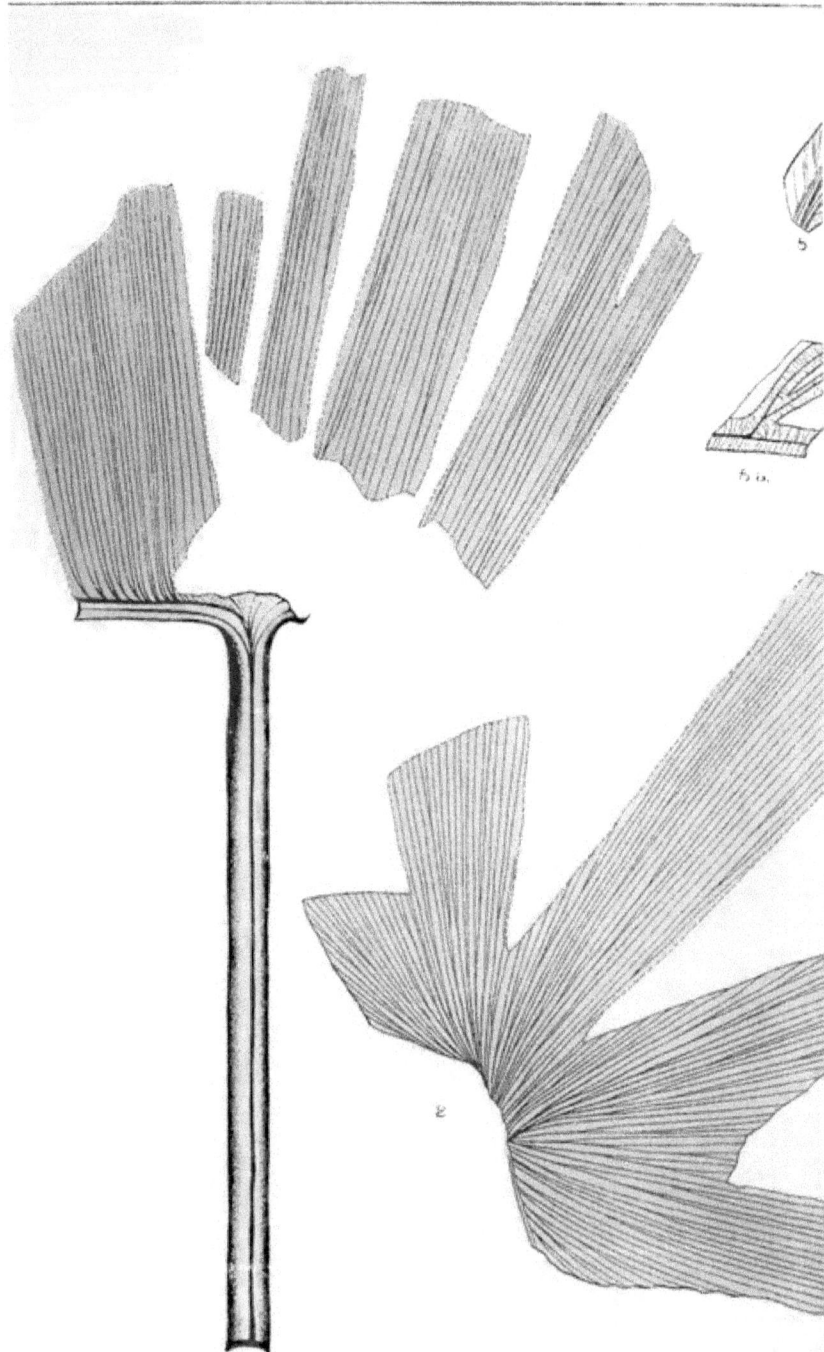

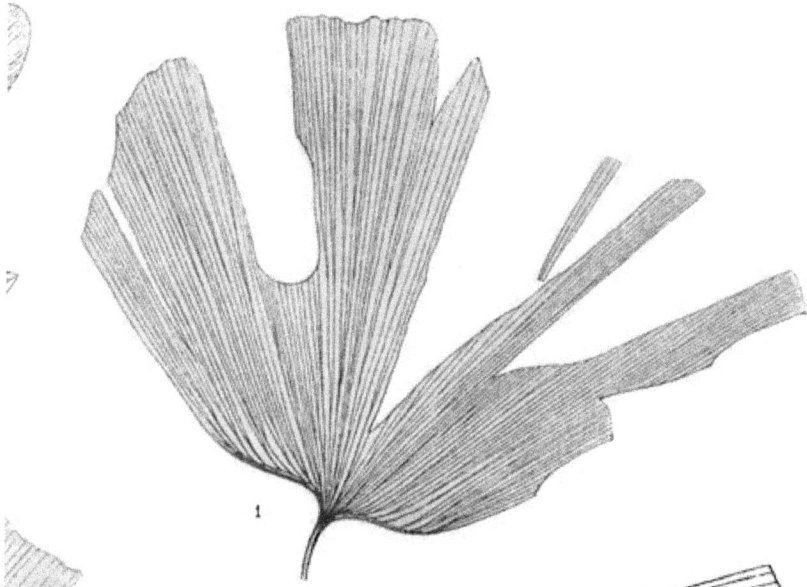

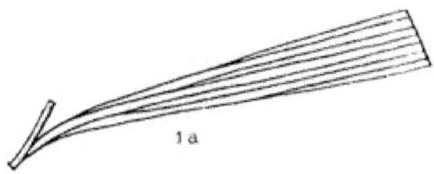

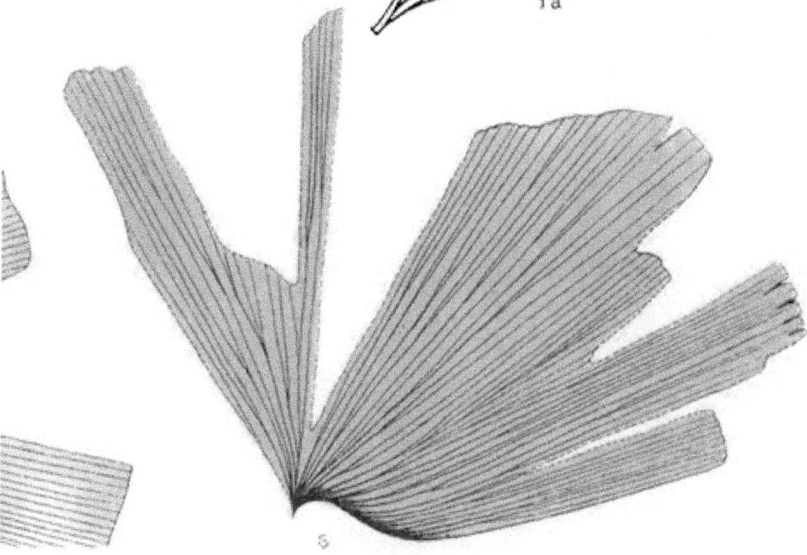